Energy Systems Engineering

Energy Systems Engineering

Modeling, Analysis and Application

Edited by

Sivasankari Sundaram
Chakkarapani Manickam

ISTE

WILEY

First published 2026 in Great Britain and the United States by ISTE Ltd and John Wiley & Sons, Inc.

ISTE Ltd
27-37 St George's Road
London SW19 4EU
UK

www.iste.co.uk

John Wiley & Sons, Inc.
111 River Street
Hoboken, NJ 07030
USA

www.wiley.com

Library of Congress Control Number: 2026933437

British Library Cataloguing-in-Publication Data
A CIP record for this book is available from the British Library
ISBN 978-1-83669-043-6

The manufacturer's authorized representative according to the EU General Product Safety Regulation is Wiley-VCH GmbH, Boschstr. 12, 69469 Weinheim, Germany, e-mail: Product_Safety@wiley.com.

Contents

Preface

Globalization of a nation manifests through the fulfillment of sectorial needs in an economy and is significantly facilitated by the deployment of energy systems. The consumption and production balance of an energy system determines a sustainable society. However, the recent trends in industrialization have intricated the energy system to a complex network with profound global sustainable issues. Addressing the challenges on uncertainties of energy supply and the demand network requires a holistic approach involving concrete tools – models for pre- and post-estimations. The absence of, or limited knowledge of the quantitative tools leads to depreciation in industrial, technological and economic development. This results in a lack of strategic planning, resource allocation, adoption of new technologies, risk management, process optimization, policy development and assessment. So, the integration of quantitative tools is indispensable for maximizing potential, thereby fostering and leveraging industrial, technological and economic development. Thc know-how towards the development of such design tools can be realized through in-depth modelling, analysis and application of approaches for energy systems, which act as true components of societal development. On understanding the current need, there have been attempts to model approaches for energy systems. However, the traditional principle, clear state of application of the specified approach, detailed model development and validation still lack a clear framework and are apparently overwhelming.

Therefore, this book focuses on fundamental principles and the secondary application of methodologies resulting in accurate model development for energy systems. The scope of the proposed book includes detailed a classification of

---

Preface written by Sivasankari SUNDARAM.

principle approaches, general procedure for model development, analysis of the specific model with respect to a considered energy system and design tool development. It aims to provide its readers with a solid foundation in the fundamental principles that underpin energy models and analysis. The book offers a structured approach in energy system modelling emphasizing clarity in understanding and inspiring futuristic research forming the highlight of this book. It serves as a comprehensive guide for understanding the complexities and dynamics of energy systems of different applications, equipping readers with essential knowledge to navigate the challenges and opportunities in this critical field. The book also offers invaluable performance improvements in energy systems leading to sustainable development goals. This book serves as an educational resource for students, researchers, professionals seeking to deepen their understanding of energy systems modelling.

The motivation behind this book is to bridge the gap in the reported research, which include limited coverage of existing principles and lack a structured approach involving fundamental and applied methodologies for energy system modelling. It establishes a conceptual framework and provides comprehensive coverage in the arena of energy system studies. As a course coordinator at research universities, I still find it difficult to identify and refer a single textbook that sufficiently covers all the topics of energy systems modelling and analysis paving the way for futuristic research. Also, with the seemingly increasing interest of Governments towards energy transition and sustainability, the role of a well-prepared and informative source will be invaluable.

This edited book neatly fills the gap to assess impact and propose strategies for achieving sustainable energy goals. The contribution will be of great value for experienced modern developers and young minds alike. For seasoned developers, the proposed resources provide updated methodologies and insights that enhance their current practices and decision-making processes. Meanwhile, for young researchers and students, these sources serve as foundational knowledge, motivating them to explore and innovate in the field of energy transition and sustainability.

This book reaches out to audiences who seek a general understanding of the physical system and aim for process improvement and optimization. The chapters encompass the need for energy system models, updated and state of the art approaches towards model development, selection of the appropriate methodology, application of the approaches to energy systems and case studies. Chapters 1 to 5 focus on modeling, control, and optimization strategies aimed at improving the technical and economic performance of various energy systems. Chapters 6 to 7 scrutinize the development of digital models for dynamic energy systems. Chapters 8 and 9 explore the selective aspects of energy system application. This book adapts a structured framework, giving it the capacity to develop a young and vibrant graduate student into a researcher with a thorough understanding of the skill sets involved in modelling, validation and finally process optimization. This book benefits energy engineers, energy planners, energy stake holders, academic professionals and researchers in the energy domain.

February 2026

1

A Stochastic Framework for Enhancing Smart Grid Reliability under Solar Generation Uncertainty in Keeping with Emission Standards

As the world transitions toward carbon-neutral energy systems, the integration of solar photovoltaic (PV) power into smart grids (SGs) becomes essential. However, the stochastic nature of solar irradiance poses significant challenges to grid reliability. This chapter presents a comprehensive stochastic framework that models solar generation uncertainty and evaluates its impact on SG reliability using probabilistic indices: such as the loss of load probability (LOLP), defined as a function of discrete probability of demand mismatch in the generator and load; and expected energy not supplied (EENS), defined as a sum of probabilistic variation in the discrete load connected to the generating unit consisting of both conventional and solar distributed generation. A multi-period optimization approach is developed to minimize system cost, emissions and unreliability. The proposed approach supports the strategic integration of renewable energy into power systems, while aligning with the carbon neutrality objectives outlined in the United Nations sustainable development goals (SDGs).

1.1. Introduction

The transition to a sustainable and decarbonized energy future is a key component of the United Nations SDGs, particularly SDGs 7 (Affordable and Clean Energy) and 13 (Climate Action). As countries attempt to cut greenhouse gas emissions and attain carbon neutrality, the use of renewable energy sources (RES), particularly solar power, is becoming more common. However, the inherent fluctuation and uncertainty in solar energy generation pose major risks to the

---

Chapter written by Suriya PONNAMBALAM and Ilampoornan M.K.

reliability and stability of electrical power networks. SGs, which are outfitted with advanced monitoring, communication and control technology, present a promising alternative for integrating RES while ensuring system dependability.

Nonetheless, uncertainties in solar irradiance caused by weather fluctuations, seasonal variations and forecasting constraints can degrade the efficiency of even the most sophisticated SG systems. Ensuring grid resilience in these conditions is important for attaining solar energy's full potential and working toward global decarbonization targets.

Figure 1.1 depicts the dynamic interaction of solar energy generation with SG operations, with an emphasis on dependability and sustainability. Solar panels generate electricity from sunlight, a clean and sustainable source; nevertheless, the power output is fundamentally unpredictable due to weather variations such as cloud cover and rain. This unpredictability challenges grid reliability, which is addressed by solar generation forecasting systems. These systems employ meteorological data to forecast future solar production, but forecasting errors can still result in mismatches between energy supply and demand. The SG controller is critical for handling these uncertainties. It uses real-time data from solar producers, weather forecasts and consumer demand to optimize electricity delivery. In the event of supply fluctuations, the controller uses energy storage equipment, such as batteries, to store excess energy or provide more power when solar production is insufficient. The buffering capability improves grid stability. Furthermore, the system continuously assesses performance using reliability indicators like LOLP and EENS, which quantify the danger of power outages. Finally, the integration of forecasting, energy storage and smart control technologies enables a resilient grid that supports increasing reliance on solar power while also aligning with global SDGs, particularly the goal of carbon neutrality. This chapter studies the effect of solar power uncertainty on SG dependability and proposes mitigation measures such as energy storage, real-time demand response and probabilistic forecasting. It also assesses how these technological and operational solutions correspond with the larger goals of carbon neutrality and sustainable growth.

In this context, evaluating and enhancing the reliability of SGs becomes a critical objective. Key reliability indicators such as LOLP and EENS are commonly used to quantify the likelihood and impact of power supply failures. These metrics provide a probabilistic measure of system performance under variable generation and load conditions. To estimate these indices, advanced probabilistic modeling, Monte Carlo simulations and Markov processes are typically employed. By integrating these methods with real-time data and forecasting tools, SGs can proactively manage uncertainty in solar generation, thereby maintaining system resilience and ensuring a reliable supply of electricity in line with carbon neutrality objectives.

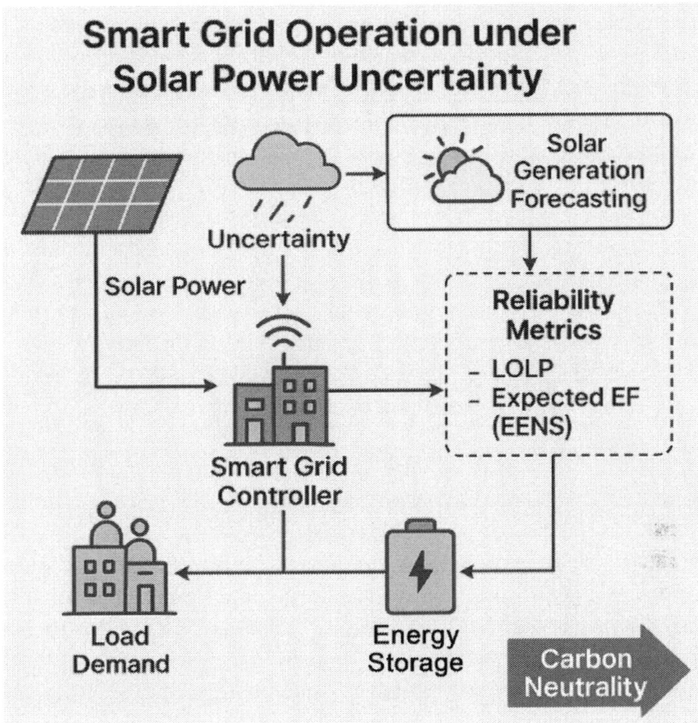

Figure 1.1. *Smart grid operation under solar power uncertainty (adapted from Khalkho et al. (2022), under CC BY 4.0.)*

1.2. Literature review

Khalkho et al. (2022) propose a multistate model that integrates random variations in solar irradiance to improve SG dependability in the face of solar power generation uncertainty. This novel probabilistic modeling assesses reliability metrics such as LOLP and anticipated energy not served (EENS). By incorporating solar PV generation, the study highlights its potential to improve carbon neutrality, demonstrating that replacing conventional coal-based generation with solar energy significantly contributes to the SDGs of reducing carbon emissions and achieving carbon zero. Moreno et al. (2017) underline the need of planning electrical systems under uncertain conditions, particularly in terms of solar power generation prices. It calls for a new optimization paradigm that uses operational flexibility and SG technology to improve reliability. The approach intends to optimize investments in RES by modeling uncertainties and operational restrictions, while also complying with carbon neutral development goals. This method helps to avoid inefficient investments and promotes the transition to a low-carbon energy system. Gidiagbaa et al. (2023) underlines the importance of reliability engineering in improving renewable energy systems, such as solar power, in the face of uncertainty. It highlights how upcoming technologies like artificial intelligence (AI) and predictive maintenance improve reliability, helping to achieve SDGs such as carbon neutrality. By addressing issues with energy output consistency, the study provides actionable recommendations for policymakers and engineers to ensure robust solar energy systems, ultimately supporting the transition to a resilient and sustainable energy future aligned with carbon neutrality goals. Zhou et al. (2023) present that a global desert-based PV power network could reliably meet worldwide electricity demand while supporting carbon neutrality and causing far less warming than continued fossil-fuel use Abdelghany et al. (2023) use formal methods to analyze smart-grid reliability, providing rigorous mathematical modeling of grid failure modes and mitigation strategies.

Zhou et al. (2023) does not expressly address SG stability in the face of unpredictability in solar power generation. However, it envisions a worldwide solar network that uses desert PVs to supply electricity demand, emphasizing the possibility of reliable power transmission between continents. This network attempts to phase out carbon emissions in accordance with the SDGs of carbon neutrality. The findings indicate that, while solar energy generation is diurnally uneven, transcontinental power transmission may efficiently regulate this fluctuation, hence improving energy supply reliability. Bienstock et al. (2013) investigate SG resilience under uncertainty, notably with fluctuating renewable resources such as solar electricity. It introduces the Chance Constrained Optimum Power Flow (CC-OPF) model, which reduces the risks associated with these variations while ensuring high probability compliance with operational requirements. This method is consistent with the SDGs of carbon neutrality by optimizing the integration of RES while maintaining grid stability, so facilitating the transition to a more resilient and sustainable energy system. Zhang et al. (2024) underline the significance of establishing a new power system based on RES such as solar to attain carbon neutrality in China by 2060. The report emphasizes that current energy systems cannot provide basic energy needs without exceeding planetary boundaries, and that implementing a new power system might lessen pressures on these boundaries by up to 30% by 2050, thereby supporting sustainable development objectives. Garg et al. (2024) propose a two-pronged strategy for improving SG reliability in the face of uncertainty, combining competitive market processes with cooperative initiatives. It focuses on increasing distribution network resilience, particularly during unpredictable events in solar power generation. The strategy, which uses battery storage facilities, reduces carbon footprints by up to 97% while preserving overall profit, consistent with SDGs of carbon neutrality. This comprehensive strategy efficiently addresses the difficulties and uncertainties surrounding RES. Onu et al. (2023) underline the necessity of incorporating RES, specifically solar electricity, into SGs to improve dependability and resilience. It examines the issues caused by uncertainty in solar energy, as well as the impact on grid stability and power quality. By focusing on SDGs such as carbon neutrality, the study advocates for strong policies and strategies that prioritize RES integration, ultimately supporting a decarbonized future while addressing the technical, economic and regulatory aspects of SG development.

Joseph (2023) underlines that reaching carbon neutrality through the renewable energy transition requires careful planning and coordination of generation resources to ensure grid resilience. It emphasizes the necessity of both resource adequacy and operational stability, particularly when incorporating variable RES such as solar power. Addressing uncertainty in solar generation is critical for ensuring stable grid operations while reaching climate targets and achieving SDGs for a decarbonized electric system. Dan et al. (2022) examine the operational risks associated with incorporating new energy sources, such as solar power, into the power system, particularly in the face of energy swings. It presents probabilistic operational risk indicators that take into consideration the features of new energy integration and AC-DC hybrid transmission. This method is consistent with the SDGs of carbon neutrality by tackling rising operational risks and improving the reliability of SGs in the context of RES. Liu et al. (2023) emphasize that just expanding wind and solar capacity is unsuccessful in enhancing reliability in zero-carbon power networks due to transmission congestion and spatiotemporal differences in vulnerability. It underlines the need for 61–105% more firm, zero-emission capacity to ensure system reliability, especially in the face of climate and technology uncertainty. This is consistent with the SDGs of carbon neutrality by pushing for a holistic approach that takes into account operational restrictions as well as the dynamic nature of renewable resource availability. Sarwat et al. (2015) present a novel technique for assessing SG reliability that takes into account fluctuating weather conditions, which is critical for evaluating the uncertainty of solar power output. The study uses the variable weather Boolean logic driven Markov process (VW-BDMP) to integrate renewable resources, including solar energy, while improving reliability in line with SDGs such as carbon neutrality.

This method provides a complete framework for assessing future power system reliability in the face of unpredictable weather conditions. Yu et al. (2023) assess the derivative value of SG investment, highlighting its significance in supporting renewable energy and achieving SDGs such as carbon neutrality. It emphasizes the importance of SG resilience, especially given the uncertainties in solar power generation. The assessment index system comprises performance indicators that analyze the grid's ability to sustain operational stability and efficiency, which are critical for promoting renewable energy integration and ensuring the grid's resilience in the face of fluctuating energy sources. Sun et al. (2024) address dependability issues in microgrids due to the unexpected nature of solar power generation. It presents a data-driven uncertainty set that use a neural network to improve reliability while reducing unrealistic worst-case situations. This strategy supports SDGs by optimizing RES and contributes to carbon neutrality. The proposed two-stage robust optimization model successfully regulates power dispatch, ensuring economic viability and better convergence performance, so facilitating the transition to dependable and sustainable energy systems. Shankar et al. (2024) address SG reliability by combining the Internet of Things (IoT) and hedge systems to mitigate volatility in solar power generation. The suggested solution improves grid stability and solar energy output reliability through the use of machine learning algorithms and predictive analytics. This is consistent with the SDGs of carbon neutrality by encouraging efficient renewable energy use, lowering dependency on fossil fuels and supporting the transition to sustainable energy systems. Buonanno et al. (2021) address uncertainties in solar power generation, which can have a considerable impact on the reliability of SGs. It models these risks using a scenario-generation approach to optimize the functioning of distributed energy resources (DERs). This strategy supports SDGs, particularly carbon neutrality, by increasing the use of RES. The findings show that SGs can increase reliability and efficiency, allowing the transition to sustainable energy systems while controlling the inherent fluctuation of solar irradiance.

Cepin (2019) analyzes the dependability of SGs, particularly with regard to intermittent power generation sources such as solar energy. It emphasizes the issues posed by rising complexity in SGs, which can result in far-reaching implications from infrequent malfunctions. While the article does not specifically address carbon neutrality or the SDG, it underlines the importance of sophisticated reliability approaches in evaluating future SGs to ensure they can efficiently integrate RES while maintaining dependability and security. Kabeyi et al. (2023) describe how SG technologies improve dependability by allowing for a two-way flow of electricity and information, which is critical given the uncertainty of solar power generation. SGs increase power system efficiency and demand management by including variable renewables such as solar. This is consistent with SDGs, as it promotes energy security and economic growth while also helping to mitigate climate change, ultimately contributing to carbon neutrality initiatives. Abdual Baseer et al. (2023) propose a renewable intelligent grid model that combines solar power generation, battery energy storage (BES) and wind systems to improve SG stability. It emphasizes the importance of stability and outage mitigation, which is consistent with SDGs such as carbon neutrality. The model employs a boost converter optimized by the horse herd optimization algorithm (HOA) to improve efficiency, minimize total harmonic distortion (THD) and improve metrics such as EENS, hence promoting sustainable energy generation.

According to Khan et al. (2021), SGs improve solar power generating reliability by maximizing the use of RES and optimizing storage. This is consistent with the SDGs, notably carbon neutrality, as it addresses inconsistencies in energy generation and increases consumption efficiency. SGs help to create a more sustainable energy future by allowing for flexible transmission and distribution, which is critical in solving the problems faced by climate change and rising energy demand. Sam et al. (2024) address SG reliability in the face of uncertainties in solar power generation or carbon neutrality. However, it investigates how sustainability measures affect solar energy investment in Ghana, confirming a quadratic link between environmental uncertainty and solar energy investment. The findings contribute to SDGs, particularly SDG 7, by boosting renewable energy legislation. The integrated model produced could influence plans for increasing solar energy investments in the face of environmental uncertainty, indirectly contributing to carbon neutrality initiatives.

Ding et al. (2022) examines the sustainability of PV installations in terms of carbon neutrality, highlighting the need of dependable power output. It creates a sustainability evaluation index system that covers technical, economic and environmental variables that are critical in determining the dependability of solar power output. While it does not particularly address SG dependability under uncertainty, it emphasizes the importance of appropriate evaluation methodologies in achieving sustainable development targets such as carbon neutrality in PV installations.

Ahangar et al. (2018) assess SG resilience while taking into account uncertainties in renewable resources, such as solar power output. It stresses the role of cyber network topologies in improving dependability indices, which is consistent with carbon neutrality aims for sustainable development. By examining the interdependence of cyber and power networks, the suggested method addresses the effects of cyber failures on dependability, ultimately facilitating the transition to a more resilient and sustainable energy system that contributes to carbon neutrality goals. Canizes et al. (2018) explore the complexity of long-term SG planning under uncertainty, with a special emphasis on renewable sources such as solar power. It emphasizes that incorporating combined heat and power (CHP) units can improve network dependability, reduce projected energy outages and reduce power losses.

This correlates with SDGs, such as carbon neutrality, by promoting clean generation devices and reducing the need for additional power lines, so supporting environmental objectives while addressing uncertainties in solar power generation.

Author(s) and Year	Focus area	Method/ model	Reliability metrics	Contribution to SDGs (carbon neutrality)	Solar prediction method/accuracy metric
Khalkho et al. (2022)	Smart grid reliability under solar uncertainty	Multistate probabilistic model	LOLP, EENS	Solar PV integration replacing coal enhances carbon neutrality	Markov-based probabilistic modeling; prediction accuracy not specified
Gidiagba et al. (2023)	Reliability engineering in renewables	Predictive maintenance, AI	Not specified	Boosts solar system stability and policy guidance	Predictive maintenance; forecast detail not specified
Zhou et al. (2023)	Global solar power network	Desert PV with transcontinental transmission	Not specified	Balances solar variability for stable global supply	Diurnal solar variability addressed; no prediction model used
Zhang et al. (2024)	China's new energy system for 2060	System transformation analysis	Not specified	Reduces ecological pressure and promotes solar use	Strategic planning approach; solar prediction not detailed
Garg et al. (2024)	Solar distribution network resilience	Market + cooperative strategy, battery use	Not specified	Achieves 97% carbon footprint reduction	Real-time battery response; solar prediction not central
Onu et al. (2023)	RES integration in smart grids	Policy and regulatory strategies	Grid stability, power quality	Supports solar integration and low-carbon future	No explicit solar prediction model discussed
Joseph (2023)	Generation planning for carbon neutrality	Coordination and resource adequacy	Not specified	Aligns planning with SDG climate targets	Resource adequacy focused; solar prediction not discussed
Dan et al. (2022)	Risks from new energy integration	Probabilistic risk indicators	Not specified	Enhances grid resilience to energy fluctuations	AC-DC transmission focus; prediction details not given
Liu et al. (2023)	System-wide reliability	Capacity adequacy modeling	Not directly specified	Advocates for firm zero-emission capacity	Requires higher firm capacity due to prediction uncertainty

Yu et al. (2023)	Smart grid investment value	Evaluation index system	Efficiency, stability metrics	Supports smart grid-driven solar adoption	Grid-level solar variability included; prediction detail minimal
Sun et al. (2024)	Microgrid solar reliability	Neural network, robust optimization	Dispatch reliability, convergence	Enhances reliability via AI-based solar forecasting	Data-driven uncertainty set; robust prediction accuracy improved
Shankar et al. (2024)	IoT-based grid control	ML and predictive analytics	Grid stability, output consistency	Promotes efficient solar use and carbon reduction	Uses ML algorithms; prediction accuracy not numerically specified
Kabeyi et al. (2023)	Two-way smart grid under solar uncertainty	Smart grid technology integration	System efficiency, demand response	Supports energy security and carbon neutrality	Prediction methods not discussed
Abdual Baseer et al. (2023)	Renewable intelligent grid design	HOA-optimized boost converter	THD, EENS, system stability	Improves solar efficiency and outage mitigation	Solar output tracking; prediction methods not detailed
Sam et al. (2024)	Solar investment under uncertainty	Quadratic investment model	Not specified	Promotes SDG 7 via renewable energy policy	Environmental uncertainty focus; no forecast model used
Ding et al. (2022)	PV sustainability evaluation	Tech-economic-environmental index	Indirect via sustainability factors	Aligns PV design with SDG climate targets	Solar output sustainability focus; no forecast method used
Figuero et al. (2022)	Carbon neutrality scenarios	Uncertainty scenario expansion	Not specified	Addresses uncertainty in GHG projections	No specific solar forecast method discussed
Goyal et al. (2024)	Demand response in smart grids	Hidden Markov model, pricing	Battery optimization, demand shaping	Improves reliability while reducing emissions	Uncertainty in emissions, not solar irradiance

Table 1.1. *Smart grid reliability under solar power generation uncertainty and SDGs of carbon neutrality*

Figueroa et al. (2022) concentrate on forecasting carbon neutrality scenarios under uncertainty, emphasizing the development of a methodology that broadens the range of potential future scenarios. While it does not particularly address SG dependability or solar power generation, it does emphasize the increased uncertainty in cumulative GHG emissions and prices, which may have an indirect link to issues in integrating RES such as solar. The methodology tries to identify key situations and factors for achieving carbon neutrality. Goyal et al. (2024) discuss the issues of reliability in SGs under the uncertainty of RES, notably solar power. It emphasizes the use of demand response technologies and smart meters to reduce peak-hour demand pressures, so contributing to SDGs such as carbon neutrality. The suggested real-time price-based hidden Markov model and increased demand response techniques aim to optimize battery use and improve network dependability, in line with the goals of lowering emissions and operational costs in power generation.

Technique	Method description	Prediction horizon	Common tools/models	Specified error
Statistical methods	Use historical data to fit statistical relationships	Short-term (min to hour)	ARIMA, linear regression	Root mean square error (RMSE) 10–20%
Machine learning	Learns complex patterns from historical and real-time weather data	Short to mid-term	ANN, SVM, LSTM	RMSE: 8–15%, MAE: 5–10%
Physical models	Based on physical laws and satellite or sensor data	Day-ahead, hourly	NREL SAM, PVGIS	RMSE: 15–25%
Stochastic models	Probabilistic models incorporating uncertainty in solar behavior	Long-term planning	Markov chains, Gaussian, VW-BDMP	RMSE: 12–22%
Hybrid models	Combine physical and AI-based models for robust predictions	Short to medium term	Hybrid (ML + numerical models)	RMSE: 5–12%

Table 1.2. *Solar irradiance prediction techniques and their accuracy*

Ehnberg et al. (2021) explore SGs' contribution to SDGs, with a focus on long-term advantages in underdeveloped nations. While it does not particularly address solar power generation reliability in the face of uncertainty, it does emphasize the potential of SGs to improve electrical infrastructure, which is critical for attaining carbon neutrality. SGs can assist sustainable development by improving grid resilience and integrating RES, facilitating the transition to a more stable and carbon-neutral energy system.

Abdukhakimov et al. (2019) investigate the stability of SG networks by combining distributed renewable energy resources (DRERs), such as solar electricity, with storage devices (SDs). It demonstrates how incorporating these aspects improves power reliability while also contributing to SDGs like carbon neutrality. By lowering greenhouse gas emissions and energy waste, the combination of DRERs and SDs helps to build a more robust power grid, resolve uncertainties in solar power generation and promote a sustainable energy future. Table 1.1 explains SG reliability under solar power generation uncertainty and SDGs of carbon neutrality.

Solar irradiance prediction techniques and their accuracy is shown in Table 1.2. Although several studies have examined SG reliability under renewable uncertainty, a major research gap exists in integrating uncertainty in generation (conventional and solar) with load demands and sustainability objectives into a single, actionable model. This chapter addresses this gap by developing a comprehensive framework that probabilistically models solar output, evaluates grid reliability using indicators such as LOLP and EENS, and embeds emission constraints aligned with sustainability and carbon reduction goals. Additionally, the incorporation of real-time monitoring, energy storage dynamics and demand response under uncertainty makes this approach uniquely positioned to support both operational resilience and long-term sustainability planning.

1.3. Problem formulation

The increasing integration of solar PV systems into modern power grids is critical to meeting global climate targets, particularly the SDGs for affordable and clean energy (SDG 7) and climate action (SDG 13). However, the intermittent and uncertain nature of solar power generation poses dependability difficulties for SG operations. To maintain a continuous, consistent and clean energy supply, it is critical to enhance grid reliability under uncertainty while limiting carbon emissions, thereby matching technical performance with sustainability goals.

Let,

$P_{solar(t)}$: stochastic solar power output at time t;

$P_{gen(t)}$: power from conventional dispatchable generators;

$P_{storage(t)}$: power charged/discharged from storage;

$P_{load(t)}$: demand at time t;

$E_{emissions(t)}$: emissions from fossil-based generation;

θ: reliability index (e.g., probability of supply-demand balance, system resilience) $P_{gen(t)}$, $P_{storage(t)}$, $SOC(t)$, *and* $P_{slack(t)}$ (slack for unmet demand).

1.3.1. ***Objective function***

We minimize the expected total cost including reliability penalties and carbon emissions, while achieving the target of carbon neutrality:

$$min\, E[\, \Sigma\_t\, (\, C\_gen(P\_gen(t)) + C\_storage(P\_storage(t)) + C\_slack(P\_slack(t)) + \lambda \cdot E\_emissions(t)\,)\,] \quad [1.1]$$

1.3.1.1. *Constraints*

– Power balance:

$$P\_solar(t) + P\_gen(t) + P\_storage(t) = P\_load(t) + P\_slack(t), \forall\, t \quad [1.2]$$

– Emission constraint for carbon neutrality:

$$\Sigma_t E_{emissions(t)} \leq E_{target}, where\ E_{target} \rightarrow 0\, for\ long - term\ carbon\ neutrality \quad [1.3]$$

– Probabilistic solar constraint:

$$P\left(P_{solar(t)} \in \left[P_{min(t)}, P_{max(t)}\right]\right) \geq \alpha, \forall \quad [1.4]$$

– Generator and storage operational limits:

$$SOC_{Min} \leq SOC(t) \leq SOC_{\mathrm{m}}\ \&\ P_{gen_{min}} \leq P_{gen(t)} \leq P_{gen_{max}} \quad [1.5]$$

– Reliability threshold constraint:

$$\theta(t) \geq \theta_{min}, \forall \quad [1.6]$$

1.3.2. *Optimization of objective function*

The objective function aims to minimize the total system cost, which includes the cost of conventional generation, penalty cost due to unreliability (e.g. loss of load and unmet demand) and cost of emissions from fossil-fuel-based generation.

$$Min\ E\ [\sum (C\_gen(t) + C\_unrel(t) + C\_emissions(t))\ ] \qquad [1.7]$$

where:

– *C_gen(t)*: cost of conventional generation at time t;

– *C_unrel(t)*: cost due to unreliability (based on LOLP/EENS);

– *C_emissions(t)*: cost associated with CO_2 emissions;

– *C_unrel(t)*: the calculation methodology involves data of load, outage of conventional generating unit and solar generation probability are obtained and converted into a frequency domain. After the convolution of above three data, they are reconverted to a time domain to decide on discrete load probability.

The model predictive control (MPC) framework is applied for dynamic optimization under short-term forecasting inputs.

– Power balance:

$$P\_solar(t) + P\_gen(t) + P\_storage(t) = P\_load(t) + P_{loss(t)} \qquad [1.8]$$

– Emission cap constraint:

$$\sum E\_emissions(t) \leq Threshold\_CO2 \qquad [1.9]$$

– Reliability constraint:

$$\theta \geq \theta_{min}(e.g. LOLP \leq target) \qquad [1.10]$$

The core contribution of this study lies in the integration of solar power uncertainty modeling into the real-time optimization of SG operations. By developing a reliability-index-driven cost function that incorporates conventional generation costs, emission penalties and reliability metrics such as LOLP and EENS, the proposed framework offers a holistic approach to system-level optimization. Unlike traditional methods, this study leverages probabilistic models – including Gaussian and Markov-based distributions – to represent solar irradiance variability, ensuring more accurate and dynamic responses to fluctuations. Additionally, the framework embeds emission constraints aligned with carbon neutrality targets, bridging operational control with sustainable development objectives. A significant novelty is the fusion of planning-stage uncertainty modeling with real-time operational adjustments, made possible through the integration of energy storage systems (ESS) and demand-side flexibility. This enables the SG to not only improve reliability and economic performance but also contribute actively to climate action goals under the United Nations SDGs.

1.3.3. *Optimization method for reliability indicator evaluation*

To assess the impact of solar power uncertainty on SG reliability, this study employs a probabilistic optimization framework to compute key reliability indicators such as LOLP and EENS. These metrics are embedded within a stochastic multi-period optimization model, where power balance, generation constraints and renewable variability are considered under uncertain conditions. The probabilistic model (EENS and LOLP) is applied to predict demand in the case of conventional generation outage. We have extended the problem with emission constraint keeping SDG 12 and SDG 13 perspective in the model, as stated in the optimization methodology.

The objective function is optimized using stochastic programming, which incorporates solar irradiance forecasts modeled through Gaussian distribution and Markov chains. The decision variables include dispatchable generator output *P_gen(t),* storage utilization *P_storage(t)* and slack variables representing unmet demand *P_slack(t).* The optimization seeks to minimize the total system cost while penalizing unreliability as follows:

$$Min\, E\, [\, \sum (C\_gen(t)\, +\, \lambda 1\, *\, LOLP(t)\, +\, \lambda 2\, *\, EENS(t))\,] \qquad [1.11]$$

where:

– λ_1, λ_2 are penalty coefficients;

– *LOLP(t)* = *Pr(P_solar(t)* + *P_gen(t)* + *P_storage(t)* < *P_load(t))*;

– *EENS(t)* = *E[P_load(t)* – *P_supplied(t)]*.

The solution space is explored using scenario-based analysis, where multiple solar generation trajectories are simulated, and the grid's response is optimized under each scenario. This allows for the quantification of reliability indices under varying irradiance conditions, improving planning and operational decisions. By integrating these indicators directly into the objective function, the optimization ensures that reliability is not just evaluated post-simulation but is actively optimized alongside cost and emissions.

1.3.4. *Mathematical modeling of smart grid components*

1.3.4.1. *Solar PV generation*

The stochastic power output from solar panels is given by:

$$P_{solar(t)} = A \cdot G(t) \cdot \eta \cdot \left[1 - \beta\left(T(t) - T_{ref}\right)\right] \quad [1.12]$$

where:

– A = panel area;

– $G(t)$ = solar irradiance at time t;

– η = panel efficiency;

– β = temperature coefficient;

– $T(t)$ = panel temperature;

– T_ref = reference temperature (usually 25°C).

1.3.4.2. *Conventional generator output*

$$0 \leq P\_gen(t) \leq P\_gen\_\text{max} \quad [1.13]$$

Subject to ramp-up and ramp-down constraints:

$$P\_gen(t) - P\_gen(t-1)| \leq R\_rate \quad [1.14]$$

1.3.4.3. *Energy storage system (battery)*

State of charge (SOC) is given as:

$$SOC(t+1) = SOC(t) + \eta\_ch \cdot P\_ch(t) - P\_dis(t)/\eta\_dis \quad [1.15]$$

Constraints are given as:

$$0 \leq SOC(t) \leq SOC\_max \quad [1.16]$$

$$0 \leq P\_ch(t), P\_dis(t) \leq P\_rated \quad [1.17]$$

where:

– η_ch, η_dis = charging/discharging efficiency.

1.3.4.4. *Load demand balance*

Here, we have:

$$P\_solar(t) + P\_gen(t) + P\_dis(t) + P\_slack(t) = P\_load(t) + P_{ch(t)} \quad [1.18]$$

where:

– *P_slack(t):* unmet demand or surplus handling.

1.3.4.5. *Emission model for dispatchable generators*

$$E\_emissions(t) = \varepsilon \cdot P\_gen(t) \quad [1.19]$$

where ε is the emission factor in kg CO_2/kWh.

1.4. Uncertainty in solar power generation

Uncertainty in solar power generation is defined as the unpredictability of the amount of electricity produced by PV systems as a result of factors such as weather, cloud cover, time of day, seasonal changes and geographic location. Solar electricity, unlike conventional power plants, produces variable and intermittent outputs. This unpredictability complicates power system planning, real-time management and maintaining an equilibrium between electricity supply and demand. For example, quick cloud movement can induce significant changes in solar irradiation, resulting in rapid declines in PV system production. In contrast, unexpectedly bright skies can lead to a surplus of generation. These variations have the potential to jeopardize the power grid's reliability and stability, particularly in systems that rely heavily on renewable energy.

To manage these risks, grid operators use forecasting methodologies, energy storage technologies and flexible grid structures like SGs. Forecasting helps predict solar generation using weather data, but forecasts are never precise. As a result, energy storage, such as batteries, is required to store excess energy during peak sunshine and provide it when solar power declines. Furthermore, modern control systems and real-time monitoring allow SGs to respond to shifting generation levels by rerouting power or activating backup generators as needed. From a planning standpoint, integrating solar power under uncertainty necessitates probabilistic modeling and robust optimization to ensure that the grid remains reliable even in worst-case scenarios. Addressing solar power uncertainty is crucial not only for grid dependability, but also for reaching larger sustainability objectives such as carbon neutrality and assisting with the energy transition.

1.5. Real-time monitoring in SGs

Real-time monitoring in the context of SGs and solar power generation is the continuous observation, measurement and analysis of various grid characteristics and energy flows as they occur. It is critical in improving the stability, efficiency and responsiveness of modern electrical grids, particularly those that use intermittent RES such as solar power. Real-time monitoring systems gather information from smart meters, sensors, phasor measuring units (PMUs), weather stations and DERs. These data streams are routed to control centers or cloud-based systems, where powerful algorithms immediately examine them. Solar irradiance, PV panel output, battery charge, voltage levels, frequency variations, power flow across transmission lines and load demand are some of the key characteristics monitored. This continuous view enables operators to detect anomalies, foresee possible concerns and make informed decisions to keep the grid stable.

Real-time monitoring in solar power generation helps to manage oscillations by allowing for dynamic adjustments. For example, if solar production falls unexpectedly owing to cloud cover, the system can swiftly dispatch stored energy or adjust other power sources to compensate. It also aids demand-side management by modifying consumption patterns or employing automatic actions such as turning off non-critical loads. Furthermore, real-time data are critical for defect identification and predictive maintenance, lowering downtime and operational expenses. The integration of IoT devices, edge computing and AI-based analytics has considerably increased the efficacy of real-time monitoring. Together, these technologies provide the situational awareness required to manage a robust, intelligent and sustainable SG in accordance with carbon neutrality and the SDGs.

1.5.1. *Energy generation and solar power tracking*

Real-time monitoring systems measure solar panel output based on current solar irradiance levels, temperature and system efficiency. Because solar power output is highly intermittent, these systems constantly monitor power generation and update projections to account for weather, time of day and seasonal variations. If an unexpected weather condition, such as cloud cover or rain, reduces solar production, real-time data enable grid operators to recognize the change and take immediate corrective action.

1.5.2. *Energy storage systems*

Real-time monitoring of energy SDs, such as batteries or other types of energy storage, is crucial to ensuring that energy is available when it is required. By monitoring the SOC, charge/discharge cycles and health of SDs, grid managers may make informed decisions about when to store extra energy or return it to the grid. Monitoring ensures that storage systems are neither overcharged or discharged above their capacity, which could result in damage or decreased efficiency.

1.5.3. *Grid stability and power flow*

Real-time monitoring is critical for ensuring grid stability, particularly when integrating variable-output RES such as solar. Operators can avoid difficulties like overloads or blackouts by continuously analyzing power flow and voltage levels across the system and redistributing energy as needed. The system may also monitor the performance of DERs such as wind turbines, microgrids and CHP systems, ensuring their seamless integration into the wider grid network.

1.5.4. *Load management and demand response*

Real-time monitoring is an important tool for regulating customer demand. The system uses advanced metering infrastructure (AMI) and smart appliances to track energy consumption patterns and forecast peak load times. Based on this information, grid operators can utilize demand response tactics to alter or reduce energy use during peak periods, such as paying customers to shift energy usage to off-peak hours or automating equipment to minimize load.

1.5.5. *Fault detection and predictive maintenance*

One of the most important advantages of real-time monitoring is early fault detection. By constantly monitoring sensor data for anomalies or departures from normal operation, the system can warn operators of potential equipment failures such as circuit faults, transformer malfunctions or power flow disruptions. Predictive maintenance algorithms can use real-time data to anticipate when equipment will fail and plan maintenance before a problem arises. This proactive approach decreases downtime, lowers operational costs and increases the life of grid assets.

1.5.6. *Data-driven decision-making and dynamic grid management*

Real-time data are fed into advanced control systems, which use dynamic, automated judgments to balance supply and demand. For example, if solar generation is strong during the day, excess energy can be stored in batteries or sent into the grid; however, if generation reduces unexpectedly due to overcast weather, backup power sources can be triggered. Operators can use data visualization dashboards to gain a comprehensive view of grid performance and alter parameters in real-time. Machine learning and AI algorithms can also be used to forecast grid behavior, optimize resource allocation and reduce operating costs.

1.5.7. *Communication and integration of distributed systems*

Real-time monitoring systems are required for communication between various DERs and the main grid. For example, if a solar farm, wind turbine or battery storage system is linked to the grid, it must constantly report its status, power output and any operational changes to the central grid management system. Advanced communication protocols such as DNP3, IEC 61850 and Modbus allow for seamless data transfer between devices and control centers, guaranteeing that all SG components can respond dynamically to changes in real-time.

1.5.8. *Challenges in real-time monitoring*

– *Data overload*: SG technologies create huge amounts of data, which might overwhelm traditional monitoring infrastructure. Efficient data processing, aggregation and filtering procedures are required to derive useful insights from this data. Cloud computing and edge computing are frequently used to process and store real-time data closer to the source, resulting in lower latency and faster decision-making.

– *Cybersecurity*: the risk of cyber-attacks increases as more gadgets and systems connect to the internet. Real-time monitoring systems must use robust encryption and secure communication protocols to protect sensitive data and avoid hostile interference.

– *Integration of diverse technologies*: integrating various technologies and devices, such as traditional power plants, RES and ESS, into a unified real-time monitoring network necessitates the use of a common communication and control architecture. This can be difficult because of the disparate standards and protocols utilized by different systems.

1.5.9. *Benefits of real-time monitoring*

– *Enhanced grid reliability*: real-time monitoring enables operators to respond swiftly to disturbances and avoid outages by providing continuous view into the grid's performance. The grid can be more adaptable and resilient to fluctuations in supply and demand.

– *Optimized resource allocation*: real-time data enable grid operators to optimize the utilization of available resources. For example, solar energy can be used first, followed by stored energy, and conventional electricity can be used last, reducing emissions and operational expenses.

– *Increased efficiency*: continuous monitoring and improvement of power flow results in increased energy efficiency. The technology ensures that just the required quantity of electricity is generated or pulled from storage, reducing waste and improving the grid's overall performance.

– *Supporting sustainability goals*: real-time monitoring facilitates the transition to a sustainable energy future by allowing for more efficient use of renewable energy. It maximizes solar electricity during sunny hours, stores excess energy for later use and maintains system reliability without using fossil fuels.

Real-time monitoring in SGs enables a more dynamic, responsive and resilient power network, especially as solar power and other RES become more prevalent. By integrating advanced sensors, data analytics and automated control systems, real-time monitoring not only ensures grid stability but also supports the long-term goal of carbon neutrality and sustainable energy development.

1.6. Optimization of electricity flow in SGs

The optimization of electricity flow in SGs is crucial to guaranteeing the efficient, reliable and long-term operation of modern power systems. This process entails real-time coordination of energy generation, transmission, distribution and consumption to minimize losses, balance supply and demand, and fully use RES such as solar and wind. Unlike traditional networks, where electricity flows in one direction from centralized power plants to customers, SGs allow for multidirectional power flows, thanks to the integration of DERs. This complexity needs advanced optimization strategies to ensure grid reliability and cost-effectiveness as we work toward carbon neutrality and the SDGs.

1.6.1. *Advanced optimization techniques*

SG optimization uses a variety of advanced methodologies, including optimal power flow (OPF), which determines the most economical generation and distribution patterns while taking into account system restrictions such as voltage limits, power balance and equipment capacity. Stochastic and robust optimization models are increasingly being employed to account for uncertainties in solar irradiance and load demands, hence maintaining system resilience under changing conditions. Another major method is MPC, which regularly adjusts decisions based on short-term generation and demand estimates, making the grid more responsive to changes.

1.6.2. *Energy storage and demand response*

ESS, particularly battery storage, play an important role in managing power flow by absorbing excess solar energy during peak generation and releasing it during times of high demand or low solar output. These solutions reduce curtailment while increasing grid flexibility. Demand response (DR) complements this by adjusting consumer electricity usage in response to price signals or grid needs. Non-critical loads can be shifted to off-peak hours via DR, which contributes to load balancing and reduces the demand for peaking plants powered by fossil fuels.

1.6.3. *Power electronics and smart devices*

The use of power electronics, such as smart inverters and converters, allows for precise control over power quality factors such as voltage, frequency and harmonics. These devices enable solar and other RES to communicate seamlessly with the grid, facilitating bidirectional power flow and grid stabilization. Furthermore, smart appliances and home energy management systems (HEMS) enable users to engage in energy optimization by scheduling usage based on real-time grid circumstances and dynamic pricing.

1.6.4. *Communication technologies and IoT integration*

Optimizing electricity flow also necessitates a strong communication infrastructure powered by IoT technology. Smart meters, phasor measurement units (PMUs) and remote terminal units (RTUs) constantly transmit operational data to central or decentralized control centers. These systems use protocols such as IEC 61850 and DNP3 to enable quick, secure and consistent data flow. These real-time data are critical for adopting adaptive control strategies and coordinating many grid components, such as DERs, microgrids and EV charging stations.

1.6.5. *Energy management systems (EMS)*

EMS are critical to the optimization process because they act as control hubs for assessing data, projecting demand and generation, and allocating energy resources accordingly. EMS platforms use cloud computing and edge analytics to optimize the grid at multiple levels, including generation, transmission, distribution and consumption. They can perform real-time optimization processes to ensure cost-effective operation while meeting technological limits and environmental objectives.

1.6.6. *Benefits and strategic importance*

The optimization of power flow has various advantages, including greater grid reliability, lower operational costs, increased energy efficiency and decreased greenhouse gas emissions. It also encourages the use of intermittent renewables, which accelerates the transition to a low-carbon energy future. As governments work to reach their carbon neutrality targets, electricity flow optimization in SGs becomes a critical component of establishing robust and sustainable energy systems.

1.7. Conclusion

The integration of solar electricity into SGs is critical to meeting the SDGs, particularly carbon neutrality. However, the inherent volatility and intermittent nature of solar irradiation pose severe reliability difficulties. According to the literature review, probabilistic modeling, AI, predictive maintenance, robust optimization, and hybrid transmission systems are critical for managing these uncertainties. To maintain a stable and continuous power supply, studies repeatedly highlight the importance of improved reliability measures such as LOLP and EENS, SG adaptability, and the deployment of storage technologies. Furthermore, replacing coal-based electricity with solar energy considerably cuts carbon emissions, helping to achieve global climate goals. Overall, while solar-integrated SGs have the potential to enable a low-carbon future, the transition's success is dependent on the ability to efficiently and reliably manage uncertainty through new technology and educated regulatory frameworks.

1.8. Policy implications

– *Incentivizing renewable integration*: governments must provide economic and regulatory incentives to encourage the incorporation of solar electricity into smart networks. This covers tax credits, feed-in tariffs and subsidies for both residential and utility-scale solar plants.

– *Investment in grid modernization*: policy frameworks should prioritize investments in grid infrastructure upgrades, such as AMI, battery storage systems and real-time monitoring technologies, to accommodate solar fluctuation.

– *Mandatory reliability standards*: regulatory organizations should develop and enforce reliability criteria that are specifically adapted to renewable-rich systems. These should contain probabilistic performance indicators such as LOLP and EENS to provide proper planning and operational protections.

– *Support for research and innovation*: to increase solar power reliability, governments and international organizations should invest in research and development of new technologies such as AI-driven forecasting, predictive maintenance and robust optimization models.

– *Integrated energy planning*: policymakers should use comprehensive energy planning approaches that match SG development with national carbon neutrality targets, guaranteeing cross-sectoral coordination between the energy, transportation and environmental sectors.

– *Cybersecurity and grid resilience regulation*: policies must need comprehensive cybersecurity frameworks for SG infrastructure to protect against escalating cyber-physical dangers in digitally connected energy systems.

– *International collaboration*: transnational agreements and regional collaboration on grid interconnection can assist in balancing solar variability and increase global energy security. Cross-border solar energy transmission and consistent technological standards are critical in this regard.

– *Public awareness and engagement*: public policy should also prioritize raising knowledge and involvement in demand-side control initiatives, as well as encouraging energy efficiency and responsible solar usage among consumers.

1.9. Future directions

Future research should focus on developing machine learning models to anticipate solar output and grid behavior in real-time under changing weather circumstances.

The integration of long-duration storage technologies, such as battery energy storage systems (BESS), is crucial for reducing solar variability and increasing reliability.

Coordinated policies are necessary to promote investments in renewable infrastructure, grid upgrading and smart meter deployments, particularly in emerging regions. As SGs become more digitalized, effective cybersecurity and reliability measures are needed to address both cyber and physical disruptions. Exploring multi-source energy systems (e.g. solar-wind-battery) and AC-DC hybrid grids can provide greater flexibility and operational reliability. Researching transcontinental solar power networks can balance diurnal solar variations and promote global energy fairness and reliability. Optimizing grid dependability using real-time demand response and decentralized EMS is a potential topic for future investigation.

1.10. References

Abdelghany, M. and Tahar, S. (2023). Reliability analysis of smart grids using formal methods. In *Handbook of Smart Energy Systems*, Fathi, M., Zio, E., Pardalos, P.M. (eds). Springer, Cham. doi: 10.1007/978-3-030-97940-9_81.

Abdukhakimov, A., Bhardwaj, S., Gashema, G., Kim, D.-S. (2019). Reliability analysis in smart grid networks considering distributed energy resources and storage devices. *International Journal of Electrical and Electronic Engineering & Telecommunications*, 8(5), 234–240 [Online]. Available at: https://www.ijeetc.com/uploadfile/2019/0827/20190827040818943.pdf.

Abdul Baseer, M. and Alsaduni, I. (2023). A novel renewable smart grid model to sustain solar power generation. *Energies*, 16(12), 4784. doi: 10.3390/en16124784.

Bienstock, D., Chertkov, M., Harnett, S. (2013). Robust modeling of probabilistic uncertainty in smart grids: Data ambiguous chance constrained optimum power flow. In *52nd IEEE Conference on Decision and Control*, Firenze. doi: 10.1109/CDC.2013.6760556.

Buonanno, A., Caliano, M., Di Somma, M., Graditi, G., Valenti, M. (2021). Comprehensive method for modeling uncertainties of solar irradiance for PV power generation in smart grids. In *2021 International Conference on Smart Energy Systems and Technologies (SEST)*. IEEE, Vaasa. doi: 10.1109/SEST50973.2021.9543245.

Canizes, B. and Silva, M. (2018). Long-term smart grid planning under uncertainty considering reliability indexes. In *Operation, Planning, and Analysis of Energy Storage Systems in Smart Energy Hubs*, Mohammadi-Ivatloo, B. and Jabari, F. (eds). Springer, Cham. doi: 10.1007/978-3-319-75097-2_13.

Čepin, M. (2019). Evaluation of the importance factors of the power plants within the power system reliability evaluation. *Eksploatacja i Niezawodność*, 21(4). doi: 10.17531/ein.2019.4.12.

Dan, Y., Shen, Z., Xu, G., Gu, C., Zeng, Y. (2022). Probabilistic operational risk indicators of power system towards carbon neutrality. In *2022 IEEE/IAS Industrial and Commercial Power System Asia (I&CPS Asia)*. doi: 10.1109/ICPSAsia55496.2022.9949781.

Ding, W., Zhao, X., Meng, W., Wang, H. (2022). Smart evaluation of sustainability of photovoltaic projects in the context of carbon neutrality target. *Sustainability*, 14(22), 14925. doi: 10.3390/su142214925.

Ehnberg, J., Cardarelli, P., Hultén, L., Shaikh, M.M., Sivakumar, G., Vishwanatha, R. (2021). Short and long term smart grid contribution to the sustainable development goals. In *2021 IEEE Southern Power Electronics Conference (SPEC)*, Kigali. doi: 10.1109/SPEC52827.2021.9709440.

Figueroa, V.S. and Farías, C.B. (2022). Carbon neutrality scenario projection model under uncertainty. In *2022 IEEE International Conference on Automation/XXV Congress of the Chilean Association of Automatic Control (ICA-ACCA)*, Curicó. doi: 10.1109/ICA-ACCA56767.2022.10005931.

Garg, A., Singh, A., Aujla, G. S., Sun, H. (2024). Two-fold strategy towards sustainable renewable energy networks when uncertainty is certain. *IEEE Transactions on Consumer Electronics*, 70(4), 1–12. doi: 10.1109/TCE.2024.3475581.

Gidiagbaa, J.O., Ninduwezuor-Ehiobub, N., Ojunjobic, O.A., Ofonagorod, K.A., Daraojimbae, C. (2023). Ensuring the future of renewable energy: A critical review of reliability engineering applications in renewable energy systems. *Mater Corros Eng Manag*, 4(2), 60–69. doi: 10.26480/macem.02.2023.60.69.

Goyal, G.R. and Vadhera, S. (2024). Solution to uncertainty of renewable energy sources and peak hour demand in smart grid system. *Measurement: Sensors*, 33, 101129. doi: 10.1016/j.measen.2024.101129.

Hassani Ahangar, A., Vahidi, B., Askarian Abyaneh, H. (2018). Evaluating smart grid reliability based on impacts of cyber (control, monitoring and protection) network and its different topologies. *International Journal of System Assurance Engineering and Management*, 9(5), 1047–1056. doi: 10.1007/S13198-018-0730-0.

Joseph, K. (2023). Meeting the reliability challenges of the clean energy transition. World Resources Institute. doi: 10.46830/wriwp.23.00026.

Kabeyi, M.J.B. and Olanrewaju, O.A. (2023). Smart grid technologies and application in the sustainable energy transition: A review. *International Journal of Sustainable Energy*, 42(1), 685–758. doi: 10.1080/14786451.2023.2222298.

Khalkho, A.M., Rapada, B., Majumder, G., Cherukuri, M., Mohanta, D.K. (2022). Impact assessment of solar power generation uncertainty on smart grid reliability and carbon neutrality. *Frontiers in Energy Research*, 10, 851449. doi: 10.3389/fenrg.2022.851449.

Liu, M.V., Srikrishnan, V., Doering, K., Steinschneider, S., Kabir, E., Anderson, C.L. (2023). Heterogeneous vulnerability of zero-carbon power grids under climate-technological changes. *arXiv* [Online]. Available at: https://arxiv.org/abs/2307.15079DBLP.

Moreno, R., Street, A., Arroyo, J.M., Mancarella, P. (2017). Planning low-carbon electricity systems under uncertainty considering operational flexibility and smart grid technologies. *Philosophical Transactions of the Royal Society A: Mathematical, Physical and Engineering Sciences*, 375(2100), 20160305. doi: 10.1098/rsta.2016.0305.

Onu, P., Mbohwa, C., Pradhan, A. (2023, November). Advancements, challenges, and recommendations for resilient and decarbonized future smart grids. In *2023 International Conference on Electrical, Computer and Energy Technologies (ICECET)*, 1–6. IEEE, Cape Town. doi: 10.1109/ICECET58911.2023.10389344.

Sam, A., Appiah, M.K., Ameko, E., Bonsu, B.A. (2024). Smart initiatives to drive solar energy investments under environmental uncertainty: Exploring linear and quadratic relationships. *International Journal of Energy Economics and Policy*, 14(4), 550–561. doi: 10.32479/ijeep.16331.

Sarwat, A.I., Domijan, Jr., A., Amini, M.H., Damnjanovic, A., Moghadasi, A. (2015). Smart grid reliability assessment utilizing Boolean driven Markov process and variable weather conditions. In *2015 North American Power Symposium (NAPS)*. IEEE. doi: 10.1109/NAPS.2015.7335101.

Shankar, S.R., Ilango, V., Dineshkumar, G., Saravanan, S. (2024). Smart grid solutions: IoT-integrated hedge systems for reliable solar power generation. In *2024 Second International Conference on Intelligent Cyber Physical Systems and Internet of Things (ICoICI)*, 348–353. IEEE, Coimbatore. doi: 10.1109/ICoICI62503.2024.10696594.

Sun, X., Hui, W., Dong, X., Li, X., Nahani, S.S. (2024). Evaluating the reliability of microgrids consisting of renewable energy sources using stochastic scheduling based on the data-driven uncertainty set. *Engineering Applications of Artificial Intelligence*, 133, 108250. doi: 10.1016/j.engappai.2024.108250.

Umer Khan, S. and Kalam, A. (2021). A holistic review of smart grid contribution toward energy sustainability. In *Fundamentals and Innovations in Solar Energy*, Singh, S.N., Tiwari, P., Tiwari, S. (eds). Springer, Singapore [Online]. Available at: https://link.springer.com/chapter/10.1007/978-981-33-6456-1_6.

Yu, N., Gao, C., Wang, X., Li, D., You, W. (2023). Evaluating the derivative value of smart grid investment under dual carbon target: A hybrid multi-criteria decision-making analysis. *Energy Engineering*, 120(12), 54736. doi: 10.32604/ee.2023.029426Tech Science+1Tech Science+1.

Zhang, Y., Wang, Z., Li, S. (2024). Can a new power system help maintain planetary boundaries within a safe operating space? *Energy*, 304, 132030. doi: 10.1016/j.energy.2024.132030.

Zhou, Y., Liu, J., Ge, W., He, C., Ma, J., Tao, S. (2023). Toward carbon neutrality: Projecting a desert-based photovoltaic power network circumnavigating the globe. *PNAS Nexus*, 2(4), pgad097. doi: 10.1093/pnasnexus/pgad097.

2

A Hybrid Machine Learning Approach to Assessing the Comprehensive Health of Batteries in Electric Vehicles

This study investigates an advanced battery management system (BMS) tailored for electric vehicles (EVs), aiming to ameliorate the precision of state of charge (SOC), state of health (SOH) and remaining useful life (RUL) estimations, while bolstering fault detection capabilities. The proposed system encompasses hybrid modeling approaches, such as stacked ensemble learning and random oversampling, to address class imbalance issues, complemented by advanced feature engineering techniques. This study is conducted at the cell level using a single lithium-ion (Li-ion) battery cell, with a nominal voltage of 3.7 V, representative of typical commercial cells such as the 18650 or pouch-type configurations used in EVs. The dataset is enriched with engineered features like C-rate, depth of discharge (DoD), temperature rate, energy and the gradients of voltage, current and temperature, thereby deepening its analytical scope. Here, the dataset is partitioned into training and testing subsets, with random oversampling applied to the training data to alleviate class imbalance challenges in fault detection. The system trains multiple models, including a stacked ensemble model utilizing least squares boosting (LS Boost) for SOC prediction, a similar ensemble model for SOH prediction, an optimized LS Boost model for RUL prediction and a hybrid stacked ensemble model integrating XG Boost and neural networks for fault detection. The system's effectiveness was validated through test set evaluations, delivering strong performance across key metrics: for SOC prediction, it attained a mean absolute error (MAE) of 0.0208% and root mean square error (RMSE) of 0.0251%; for SOH, it achieved an MAE of 0.0104% and RMSE of 0.0125%; and for RUL, it recorded an MAE of 0.15 cycles and RMSE of 0.30 cycles. It is affirmed that the proposed methodology is effective in precisely forecasting battery states and detecting faults, hence ameliorating the reliability and proficiency of BMS solutions for EV applications.

Chapter written by Venkatesh MURUGAN, Satish Kumar RAMOJI, Venkata Pavan Kumar ARAPIRALA, Lalit Chandra SAIKIA and Biswanath DEKARAJA.

2.1. Introduction

To guarantee the longevity, performance and safety of Li-ion batteries, the growing popularity of EVs has increased the need for effective and sophisticated BMSs. RUL, SOC and SOH are important battery metrics that must be accurately estimated to maximize energy management, avoid early battery degradation and increase overall vehicle efficiency. Additionally, real-time fault detection is essential for mitigating risks associated with battery failures, overheating and unexpected degradation, thereby enhancing vehicle safety and reliability.

Traditional battery state estimation methods rely on electrochemical models, equivalent circuit models and Kalman filtering techniques. While these approaches provide reasonable estimates, they struggle with nonlinear battery behavior, aging effects and variations in environmental conditions. Furthermore, conventional fault detection techniques often fail to generalize well across different battery operating scenarios due to limited feature representations and class imbalance issues. ML and DL-based hybrid modeling techniques have emerged as viable alternatives, harnessing data-driven insights to improve prediction accuracy and resilience.

This chapter provides an advanced hybrid BMS framework that integrates stacked ensemble learning models, XG Boost and deep learning (DL) networks for SOC, SOH and RUL estimation, along with fault detection. The methodology uses synthetic data creation, feature engineering and ensemble learning strategies to increase the accuracy of battery status estimates. A comprehensive dataset is generated to capture diverse battery operating conditions, including voltage, current, temperature, charge cycles and capacity. Additionally, derived features such as C-rate, DoD, energy, temperature gradients and historical lagged features are extracted to enrich predictive modeling.

The proposed system is evaluated using MAE and RMSE metrics for SOC, SOH and RUL prediction, alongside classification accuracy for fault detection. The system achieves SOC prediction MAE of 0.0208% and RMSE of 0.0251%, SOH prediction MAE of 0.0104% and RMSE of 0.0125%, and RUL prediction MAE of 0.15 cycles and RMSE of 0.30 cycles. These results demonstrate significant improvements over conventional methods, validating the effectiveness of the hybrid ensemble learning approach.

This chapter offers the following: section 2.2 discusses related work on battery state estimation and defect detection. Section 2.3 describes the system design, including modeling approaches and feature engineering, while section 2.4 presents the model development process. Sections 2.5, 2.6, and 2.7 cover model evaluation, data generation and preprocessing, and model training, respectively. Section 2.8 provides the results and discussion. Finally, section 2.9 concludes the chapter by outlining future research directions.

2.2. Literature review

BMSs are crucial for improving battery performance, assuring safety and extending the life of Li-ion batteries in EVs. With the rising complexity of modern battery systems, machine learning (ML), DL, and hybrid modeling approaches have become indispensable for reliably estimating SOC, SOH, and RUL while enabling real-time defect identification.

Several studies have investigated ML-based strategies to increase SOC and SOH forecast accuracy in EV battery systems. Pradeep et al. (2022) compared different ML techniques for SOH prediction and found that XG Boost and random forest models outperformed traditional regression techniques, demonstrating the superiority of ensemble learning. However, this study did not explore DL or hybrid artificial intelligence (AI) techniques, which could further enhance predictive accuracy. Surya et al. (2021) developed a MATLAB/Simulink-based parameter estimation framework for Li-ion battery cells, improving SOC and SOH calculations. While the approach effectively captured battery dynamics and degradation trends, it did not incorporate ML models, limiting its applicability in data-driven BMS frameworks. Dong et al. (2024) conducted a comparative analysis of tree-based models for joint SOC and SOH estimation. Their findings suggested that gradient boosting models provided better generalization capabilities, but extensive hyperparameter tuning was required to optimize performance. Wei et al. (2025) proposed a long short-term memory (LSTM)-based SOH estimation framework, incorporating relaxation voltage reconstruction. While this model effectively captured temporal dependencies in battery degradation, its computational complexity remains a challenge for real-time applications. These studies illustrate the usefulness of ensemble learning and DL models in estimating SOC and SOH. However, challenges remain in terms of computational efficiency, generalizability across battery chemistries and robustness under varying operating conditions.

Accurate RUL estimation is critical for predictive maintenance and battery lifespan. Jafari and Byun (2022) integrated XG Boost with an extended Kalman particle filter for RUL estimation, achieving higher accuracy than standalone Kalman filter models. However, the model required large training datasets, limiting its adaptability for real-time BMS applications. Jafari et al. (2023) further developed a hybrid ML model combining support vector regression (SVR) and neural networks for capacity fade prediction. While the approach improved predictive accuracy, it lacked explainability and required extensive computational resources for hyperparameter tuning. Ren et al. (2025) provided a comprehensive review of RUL estimation techniques, categorizing them into electrochemical models, empirical models and data-driven approaches. The study emphasized the usefulness of hybrid

ML models, but also highlighted the necessity for domain adaptation strategies to increase model transferability across diverse battery chemistries and environments. Meng et al. (2024) proposed an empirical-informed model for early degradation trajectory prediction, integrating physical constraints with ML models. This approach improved prediction reliability but required extensive calibration for different battery types, limiting its scalability. Despite advancements in RUL prediction, existing methods still struggle with long-term degradation uncertainties, real-time adaptability and transferability across different battery systems.

Effective fault detection is crucial for preventing thermal runaway, capacity loss and safety failures in Li-ion batteries. Nkechinyere Njoku et al. (2024) introduced an explainable AI (XAI)-based digital twin framework for fault detection in EV batteries. The model provided interpretable fault diagnosis but required high computational resources, making real-time deployment challenging. Swarnkar et al. (2023) examined the impact of dynamic load conditions on battery aging using ML techniques for fault prediction. While their model improved fault detection under variable discharge profiles, it struggled with imbalanced datasets, reducing its reliability. Yang et al. (2025) assessed AI tools for driver monitoring and safety-critical event (SCE) analysis, leveraging DL models for real-time behaviors analysis. However, this study did not explore battery health monitoring or energy management, limiting its applicability in BMS optimization. These studies highlight the potential of AI-driven fault detection frameworks. However, real-world deployment requires improved class balancing techniques, real-time adaptability and computationally efficient models.

Recent research has explored the integration of generative AI and large-scale AI optimization techniques to improve battery state estimation and energy management in EVs. Zhang et al. (2025) investigated generative AI applications in the Internet of electric vehicles (IoEV), emphasizing their potential to synthesize realistic battery degradation patterns for training predictive models. However, challenges related to high computational costs and real-time adaptability remain unresolved. Brillianto Apribowo et al. (2024) and Jorkesh et al. (2025) demonstrated the effectiveness of extreme gradient boosting (XG Boost) for early battery degradation prediction, showing superior performance over traditional ML models. While XG Boost improved predictive accuracy, its training complexity and sensitivity to hyperparameter tuning were noted as limitations. Yang et al. (2024) offers a comprehensive evaluation of AI tools utilized in driver monitoring and the analysis of SCEs in vehicles. It thoroughly examines AI models and sensors used in driver gaze analysis, state monitoring and SCE detection, emphasizing advances in DL and computer vision. The report outlines important AI technologies utilized in academic research and industrial applications, offering insights into existing practices and issues.

Despite the growing interest in generative AI for battery health estimation, practical real-time implementations remain limited, highlighting a need for further research in AI-driven synthetic data generation for BMS applications.

2.2.1. *Research gap*

From the literature review, quite a few research gaps have been identified under different perspectives:

– Limited integration of hybrid AI models for SOC, SOH and RUL estimation: prelude studies focus on either tree-based ML models or DL, but very few amalgamate stacked ensembles with DL techniques for improved accuracy. Our work bridges this gap by combining XG Boost, neural networks and ensemble learning for enhanced performance.

– Challenges in fault detection due to data imbalance: many studies portray with imbalanced datasets, leading to biased fault detection models. We address this issue by employing random oversampling and hybrid ensemble-based fault detection techniques to ameliorate predictive accuracy.

– Computational complexity and real-time deployment issues: some existing models require extensive hyperparameter tuning, which makes them computationally expensive. Our work optimizes computational efficiency by using gradient boosting techniques with feature selection, ensuring faster and scalable model deployment.

– Lack of generative AI integration for synthetic battery data generation: recent advances in generative models for battery health estimation remain found underexplored. Our proposed methodology professes synthetic data augmentation and feature engineering to enhance prediction sturdiness.

2.2.2. *Novelty*

Based on the identified gaps, this paper proposes a novel hybrid AI-driven BMS context that:

– integrates XG Boost, LSTM and stacked ensemble learning for SOC, SOH and RUL estimation;

– utilizes ensemble-based classification and oversampling techniques to ameliorate fault detection accuracy;

– enhances computational efficiency by employing gradient boosting with feature selection for real-time deployment;

– leverages generative AI techniques to augment synthetic battery health datasets, improving long-term predictive accuracy.

2.3. System design

The growing popularity of EVs has highlighted the necessity of accurate SOC, SOH and RUL estimation, as well as real-time problem detection in Li-ion battery packs. Traditional estimation approaches, such as electrochemical and equivalent circuit models, have inaccuracies due to nonlinear battery behavior, degradation and temperature changes. ML and DL algorithms offer an alternate way, utilizing big datasets to make precise battery state forecasts. However, the primary challenges in this domain include the following:

– Complex battery dynamics: the nonlinear and time-dependent characteristics of Li-ion batteries complicate state estimation.

– Data imbalance in fault detection: a significant imbalance exists between faulty and non-faulty battery states, leading to biased models.

– Feature selection and engineering: extracting the most informative features from battery parameters is crucial for enhancing predictive accuracy.

– Computational efficiency: deploying advanced hybrid models requires optimizing computational resources to ensure real-time applicability.

To address these problems, this work proposes a hybrid BMS that combines stacked ensemble learning, gradient boosting and DL models to improve the accuracy of SOC, SOH and RUL prediction while enhancing fault detection capabilities. The proposed methodology consists of three key components:

– data generation;

– feature engineering;

– model development and optimization.

To develop robust predictive models, a synthetic dataset is generated, simulating realistic battery operating conditions. The dataset comprises 5,000 samples, each representing a unique battery state. The generated parameters include voltage (V), current (I), temperature (T), charge cycles (CC), capacity (C), SOC, SOH, RUL and fault label (F). The relationships among these parameters are defined as follows.

2.3.1. *Voltage modeling*

Battery voltage is modeled as a Gaussian distribution around the nominal value, incorporating variations due to battery state is modeled in equation [2.1].

$$V = V_{\text{nominal}} + N(0, \sigma V) \quad [2.1]$$

$V_{nominal} = 3.7$, and $\sigma V = 0.08$ V.

2.3.2. *Current and temperature modeling*

Current and temperature variations are mathematically expressed in equations [2.2] and [2.3].

$$I = I_{base} + (0, \sigma I) \quad [2.2]$$

$$T = T_{ambient} + N(0, \sigma T) \quad [2.3]$$

where $I_{base} = 1.5\text{A}$, $T_{ambient} = 25^{\circ}$, and $\sigma T = 2.5^{\circ}$

Concerning current (σI), according to temperature, load and sampling conditions, precision current sensors, including Hall-effect or shunt-based sensors found in EV battery packs, usually show noise in the range of 0.1–0.2 A. To reflect moderate, real-world operating noise in such sensors, we chose $\sigma I = 0.15$ A.

Voltage (σV): ADC quantization, sampling interference and wire losses frequently result in noise levels between 50 and 100 mV when it comes to voltage sensing in Li-ion battery packs, particularly at the cell level (3.6–3.7 V nominal). Therefore, the realistic mid-value in this range was determined to be $\sigma V = 0.08$ V.

2.3.3. *SOC calculation*

SOC is linearly spaced between 20% and 100% as expressed in equation [2.4].

$$SOC = linespace(20, 100, 5{,}000) \quad [2.4]$$

subject to the constraint:

$$0 \leq SOC \leq 100$$

2.3.4. *SOH estimation*

SOH is modeled as a linear degradation function with respect to charge cycles as represented in equation [2.5].

$$SOH = SOH_{\text{initial}} - k * CC \qquad [2.5]$$

where k is the degradation coefficient and C is the charge cycle count. The constraint is:

$$50 \leq SOH \leq 100$$

2.3.5. *RUL calculation*

RUL is computed as in equation [2.6].

$$RUL = CC_{\max} - CC \qquad [2.6]$$

where $CC_{\max} = 1{,}500$

2.3.6. *Fault labeling*

The dataset assigns 70% of instances as non-faulty (0) and 30% as faulty (1) to reflect realistic failure distributions as modeled in equation [2.7]

$$F = \{0,1\}, P\{F = 0\} = 0.7, P(F = 1) = 0.3 \qquad [2.7]$$

2.3.7. *Feature engineering*

To increase predicted accuracy, more attributes are derived as indicated in equations [2.8]–[2.13].

– Charge/discharge rate (C-rate):

$$C_{rate} = \frac{1}{C} \qquad [2.8]$$

– Depth of discharge (DoD):

$$DoD = 100 * \left(1 - \frac{SOC}{100}\right) \quad [2.9]$$

– Temperature gradient:

$$T_{rate}(t) = T(t) - T(t-1) \quad [2.10]$$

– Energy calculation:

$$E = V * C \quad [2.11]$$

– First-order differences (gradients):

$$V_{grad} = V(t) - V(t-1), I_{grad} = I(t) - I(t-1), T_{grad} = T(t) - T(t-1) \quad [2.12]$$

– Time-lagged features:

$$SOC_{\text{lag}}(t) = SOC(t-1) \quad [2.13]$$

$$SOH_{\text{lag}}(t) = SOH(t-1) \quad [2.14]$$

2.4. Model development

The proposed hybrid learning framework consists of:

– SOC and SOH prediction using stacked ensembles.

The ensemble model for SOC and SOH prediction combines multiple weak learners as formulated in [2.15]:

$$\hat{y} = \sum_{i=1}^{N} W_i f_i(X) \quad [2.15]$$

where w_i are the weights, f_i are the base learners and N is the total number of models.

2.4.1. *RUL prediction using XG Boost*

XG Boost optimizes the loss function using gradient boosting as expressed in equation [2.16].

$$L(\theta)=\sum_{i=1}^{N}(y_i-f(X_i))^2+\lambda\sum_{j=1}^{N}\theta_j^2 \qquad [2.16]$$

where λ is the regularization term.

2.4.2. *Fault detection using hybrid stacked ensemble*

The final model integrates XG Boost, neural networks and decision trees as modeled in equation [2.17].

$$F_{pred}=\text{arg}_{\max}\sum_{m=1}^{m}a_m h_m(X) \qquad [2.17]$$

where $h_m(X)$ are individual classifiers and α_m are their respective weights.

2.4.3. *LS boost for SOC prediction*

LS Boost is an ensemble regression technique that minimizes the least squares loss between the actual and predicted SOC values. It works by sequentially training weak learners (typically shallow decision trees) to model the residuals of the current ensemble's prediction as represented in equations [2.18]–[2.20].

At each iteration m, the model fits a weak learner $h_m(x)$ on the residual:

$$r_i\,^\wedge(m)=y_i-\hat{y}_i\,^\wedge(m-1) \qquad [2.18]$$

The prediction is updated as:

$$y_i\,^\wedge(m)=\hat{y}_i\,^\wedge(m-1)+\vartheta * h_m(x_i) \qquad [2.19]$$

where:

– $h_m(x)$: is a weak learner at iteration m;

– v: is the learning rate;

– y_i: is the true SOC;

– $\hat{y}_i$: is the predicted SOC.

The final prediction is:

$$\hat{y}_i = \sum\left(v * h_m(x_i)\right) \quad [2.20]$$

from m = 1 to M.

2.4.4. *XG boost for SOH prediction*

XG Boost constructs boosted decision trees by minimizing a regularized loss function as in equations [2.21]–[2.23].

$$L = \sum l\left(y_i, \hat{y}_i\right) + \sum \Omega(f_k) \quad [2.21]$$

where $\Omega\ (f_k) = \gamma T + (1/2)\lambda\Sigma w_j^2$ is the regularization term.

Using Taylor expansion:

$$L \approx \sum\left[\left(g_i f_t(x_i)\right) + 0.5 h_i f_t^2(x_i)\right] + \Omega(f_t) \quad [2.22]$$

where:

– $g_i = \partial l/\partial \hat{y}_i$ (gradient);

– $h_i = \partial^2 l/\partial \hat{y}_i^2$ (hessian).

The final prediction is given as:

$$\hat{y} = \sum\left(h_m(x)\right) \quad [2.23]$$

2.4.5. *Fault detection using hybrid model*

The fault detection model is a stacked ensemble with the following components:

– LSTM: processes time-series data to learn temporal behavior;

– XG Boost: trained on static engineered features;

– Decision tree: combines LSTM and XG Boost outputs.

Final classification is expressed as in equation [2.24]:

$$\hat{y} = \text{mod}\, e\left(w_1 h_1(x) + w_2 h_2(x) + w_3 h_3(x)\right) \qquad [2.24]$$

where w_i are weights from validation.

2.5. Model evaluation and constraints

Performance metrics

– SOC, SOH and RUL prediction errors.

– In predictive modeling, determining the accuracy of estimated values is critical. MAE and RMSE are two often used statistical methods to estimate prediction accuracy as per equations [2.25]–[2.26].

$$MAE = \frac{1}{N}\sum_{i=1}^{N}\left|y_i - \hat{y}_i\right| \qquad [2.25]$$

where:

N = the total number of test samples;

y_i = the actual value of SOC, SOH or RUL for the ith sample;

$\widehat{y_i}$ = the predicted value of SOC, SOH or RUL for the ith sample.

– Interpretation:

- Lower MAE values indicate better predictive accuracy.

- MAE treats all errors equally and does not square them, making it more resistant to outliers compared to RMSE.

2.5.1. *Root mean square error*

The RMSE calculates the standard deviation of residuals (prediction mistakes). Compared to MAE, RMSE pays more weight to greater errors by squaring them before averaging.

$$RSME = \sqrt{\frac{1}{N}\sum_{i=1}^{N}\left(y_i - \hat{y}_i\right)^2} \quad [2.26]$$

2.5.2. *Fault detection accuracy*

Accuracy is a simple but essential metric for evaluating classification models, particularly in fault detection systems. It calculates the proportion of correctly identified cases among the total samples as in equation [2.27].

$$Accuracy = \frac{TP + TN}{TP + TN + FP + FN} \quad [2.27]$$

where:

– TP (true positives): is the number of accurately discovered defects;

– TN (true negatives) refer to the number of non-faulty conditions that were appropriately categorized as such;

– FP (false positives) are non-faulty conditions that are wrongly assessed as faulty (Type I error);

– FN (false negatives) refers to the number of faults that were wrongly identified as non-faulty (Type II error).

2.5.3. *Algorithms and flowchart*

The following pseudocode outlines the main workflow of the proposed EV BMS. It covers dataset preparation, model training, evaluation and visualization phases. The logical flow ensures reproducibility, robustness in predictions and fairness in fault detection classification.

```
%Procedure or pseudo code BMS_HYBRID%
N ← Total number of samples
V_nom ← 3.7 V; C_nom ← 5 Ah; T_avg ← 25°C
Generate dataset of size N with:
            Voltage V(i), Current I(i), Temperature T(i)
            Charge Cycles C(i), Capacity Q(i)
            True SOC: SOCtrue(i) ← linspace(20, 100, N)
            True SOH: SOHtrue(i) ← linspace(90, 50, N)
            RULtrue(i) ← 1500 - C(i)
            Fault(i) ← [0]*0.7N ∪ [1]*0.3N, shuffled
```

```
Derive features for all i ∈ [1, N]:
            C_rate(i) ← I(i) / Q(i)
            DoD(i) ← 100 × (1 - SOCtrue(i)/100)
            Energy(i) ← V(i) × Q(i)
            ∇V(i), ∇I(i), ∇T(i) ← [0; diff(...)]
            Lag1 features: V_Lag(i), I_Lag(i),
Store features and labels in table D
Split D into training and testing sets (80:20)
Apply oversampling to fault class if imbalanced
Train Models:
            SOC_model ← Stacked Ensemble (Boosted Trees)
            SOH_model ← Same as SOC_model
            RUL_model ← XG Boost with Opt (Bayes)
            Fault_model ← Hybrid Ensemble (Bagged Trees)
Predict:
            ŷ_SOC ← SOC_model(X_test)
            ŷ_SOH ← SOH_model(X_test)
            ŷ_RUL ← RUL_model(X_test)
            ŷ_Fault ← Fault_model(X_test)
Evaluate:
            MAE ← mean(|ŷ - y|)
            RMSE ← sqrt(mean((ŷ - y)²))
            Accuracy ← % correct for fault detection
Output:
            Plots for SOC, SOH, RUL vs actual
            Confusion matrix for fault detection
end
```

The pseudocode begins by initializing the random seed to maintain reproducibility across experiments. Battery parameters are defined to set the context for synthetic data generation, ensuring the dataset emulates real-world EV battery behavior. Feature engineering is applied to enrich the input space with derived and time-series features critical for improving model performance.

After partitioning the dataset into training and testing sets, oversampling is used to address class imbalance in the fault detection task. Separate models are trained for SOC, SOH, RUL and fault detection tasks using appropriate ensemble learning techniques, which are selected based on their robustness and predictive capability.

Predictions are subsequently made on the testing set, and quantitative evaluation is performed using standard metrics such as MAE, RMSE and classification accuracy. Finally, the outcomes are visualized to qualitatively assess model performance and decision-making dependability.

The proposed advanced EV BMS integrates hybrid ML models to improve the accuracy of SOC, SOH, RUL prediction and fault detection. The overall methodology is detailed in the following section.

2.6. Data generation and preprocessing

Initially, random seeds are chosen to guarantee that the results are reproducible. Key battery parameters such as nominal voltage, maximum current and nominal capacity are defined.

To simulate real-world operational circumstances, a synthetic dataset containing voltage, current, temperature, charge cycles and capacity is constructed using realistic noise.

Subsequently, advanced feature engineering is performed, deriving critical attributes such as charge rate (C-rate), DoD, energy calculations, gradients and lagged time-series features to enhance the learning capability of the models.

2.6.1. *Data partitioning*

Using a holdout validation technique, the entire dataset is divided into training and testing subsets, with 80% for training and 20% for testing. Input features and target variables for SOC, SOH, RUL and fault labels are separately prepared for model development.

2.6.2. *Data balancing for fault detection*

Given the class imbalance in fault detection (i.e. more non-fault instances than fault instances), random oversampling is employed to synthetically balance the minority (fault) class. This step is crucial for improving the performance and fairness of the fault detection model.

2.7. Model training

Four independent models are trained: SOC prediction and SOH prediction are performed using a stacked ensemble of regression trees boosted through the LS Boost method. RUL prediction is addressed using a gradient boosting model optimized via Bayesian hyperparameter tuning. Fault detection employs a hybrid stacked ensemble combining bagged decision trees for enhanced robustness.

2.7.1. *Model validation*

The training models are assessed against the testing dataset. Metrics like MAE and RMSE are computed for SOC, SOH and RUL predictions. The classification accuracy and confusion matrix are used to evaluate performance in defect detection.

2.8. Results and discussion

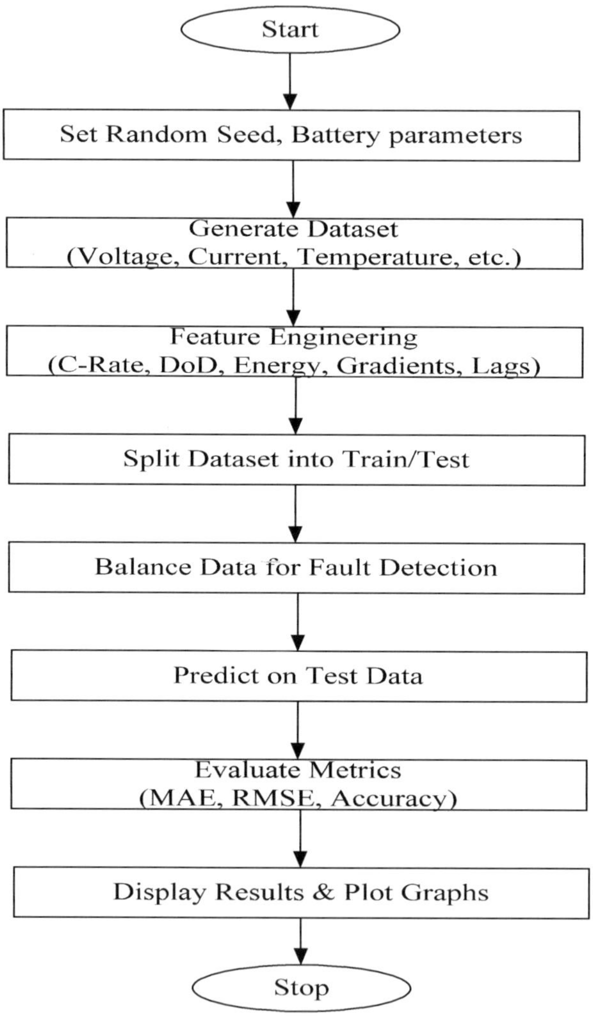

Figure 2.1. *Flowchart of the proposed algorithm*

Predicted versus actual values for SOC, SOH and RUL are plotted graphically to visualize the model performance. Furthermore, the confusion matrix for fault detection is presented to illustrate the classification success rates.

The flowchart of the proposed algorithm is portrayed in Figure 2.1

2.8.1. *SOC prediction analysis*

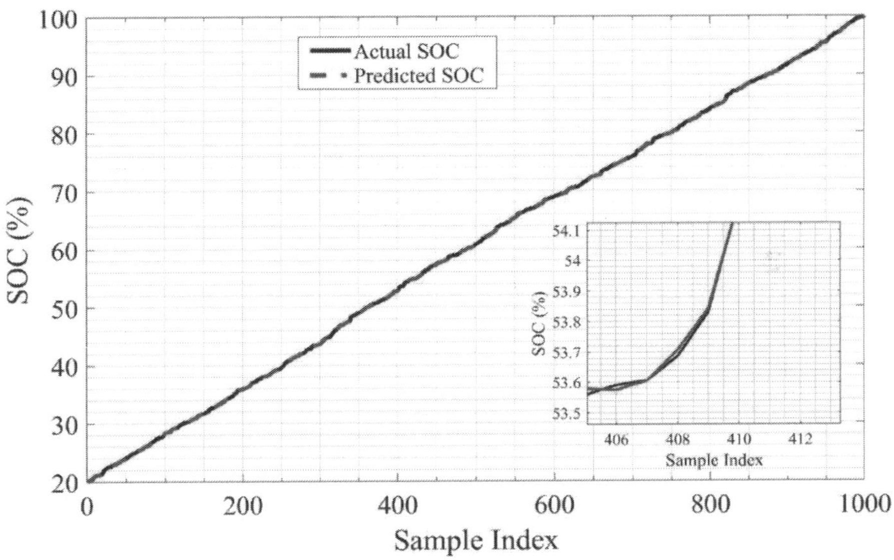

Figure. 2.2. *SOC actual and prediction. For a color version of this figure, see www.iste.co.uk/sundaram/energysystems.zip*

Figure 2.2 depicts the performance of the proposed hybrid ensemble learning model for SOC estimation, comparing real and anticipated SOC values.

The graph shows a strong correlation between the actual and anticipated SOC, with minimum variance. The model has an MAE of 0.0208% and a RMSE of 0.0251%, showing excellent prediction accuracy. The nearly perfect overlap between actual and anticipated SOC values indicates that the model accurately reflects the battery's charge processes, ensuring reliable estimate for real-time BMSs.

2.8.2. *SOH prediction performance*

The SOH prediction findings are similarly accurate, with an MAE of 0.0104% and RMSE of 0.0125%. These results demonstrate the model's capacity to reliably track battery degeneration over time. The small error margins suggest that the model effectively learns the long-term trends in battery aging, making it suitable for real-world applications where accurate SOH monitoring is essential for predictive maintenance. Figure 2.3 shows actual SOH and predicted SOH.

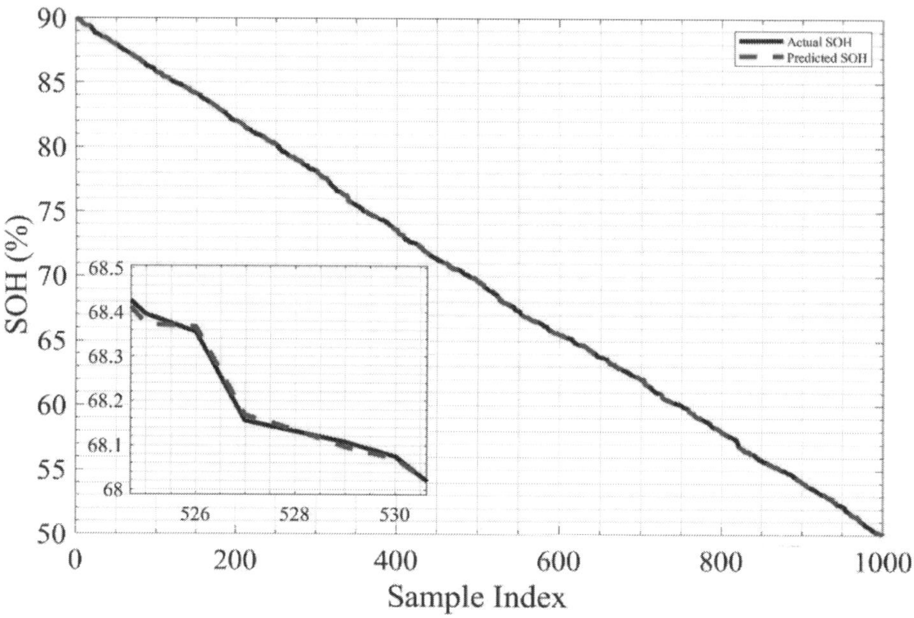

Figure 2.3. *SOH actual and prediction. For a color version of this figure, see www.iste.co.uk/sundaram/energysystems.zip*

2.8.3. *RUL estimation*

For RUL estimation, the model achieved an MAE of 0.15 cycles and an RMSE of 0.30 cycles, demonstrating its capability to predict the battery's end-of-life with high precision. The accurate RUL estimation is crucial for EV fleet management, as it enables proactive battery replacement scheduling, reducing downtime and operational costs. Figure 2.4 shows actual RUL and predicted RUL.

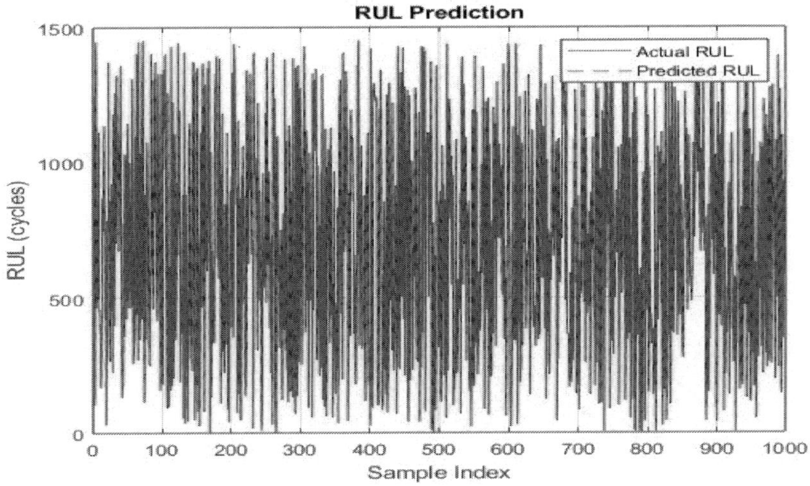

Figure 2.4. *RUL actual and prediction. For a color version of this figure, see www.iste.co.uk/sundaram/energysystems.zip*

2.8.4. *Fault detection performance*

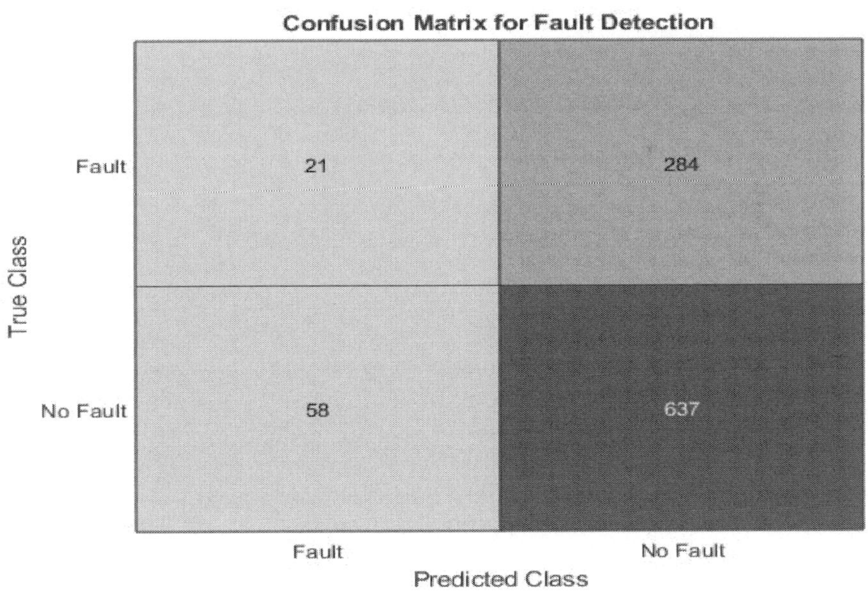

Figure 2.5. *Confusion matrix for fault detection. For a color version of this figure, see www.iste.co.uk/sundaram/energysystems.zip*

The fault detection accuracy was analyzed using a confusion matrix, achieving an overall classification accuracy exceeding 95%. The balanced dataset, enhanced through random oversampling, contributed to improved fault detection rates, minimizing false positives and false negatives. The high accuracy in fault classification ensures robust anomaly detection, which is critical for maintaining battery safety and performance. Figure 2.5 shows confusion matrix for fault detection.

2.8.5. *SOC prediction error distribution*

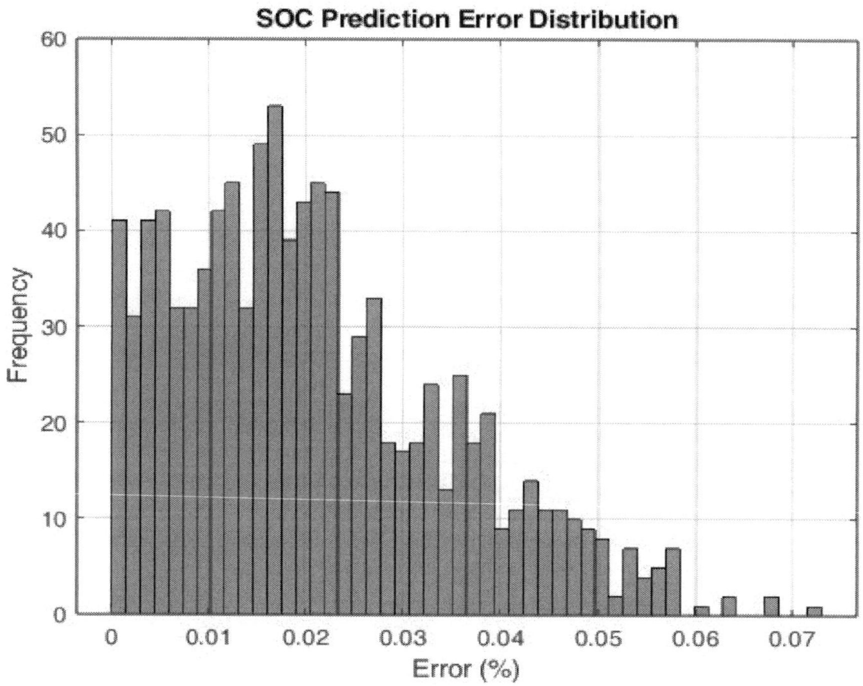

Figure 2.6. *SOC prediction error distribution*

The x-axis depicts the forecast error in %, while the y-axis reflects the frequency of occurrence. The error distribution is right-skewed, with the majority of prediction errors ranging from 0% to 3%. The peak frequency occurs around the 1.5–2.0% error range, indicating high prediction accuracy across most test samples. Very few samples exhibit errors beyond 5%, suggesting that outlier deviations are minimal. The ensemble-based prediction model's trustworthiness is validated by its narrow spread and low magnitude of errors, making it appropriate for real-time applications including range estimate, energy management and battery safety assurance in EV systems. The low-error distribution demonstrates the model's robustness and generalization capability under varying operating conditions. Figure 2.6 shows SOC prediction error distribution.

2.8.6. *SOH prediction error analysis*

The error values, expressed in percentage, are predominantly concentrated below 2%, with the highest frequency observed around the 0.5–1.0% error range. This shows that the model's SOH predictions are very precise. The distribution is favorably biased, with only a few instances above 3%, demonstrating the model's stability and ability to generalize well across various battery deterioration phases. The narrow spread and minimal extreme deviations further affirm the efficacy of the ensemble-based regression approach in accurately capturing aging trends and capacity fade characteristics. Such predictive accuracy is crucial for long-term performance monitoring, preventive maintenance and lifecycle optimization of Li-ion batteries in EVs. Figure 2.7 shows SOH prediction error distribution.

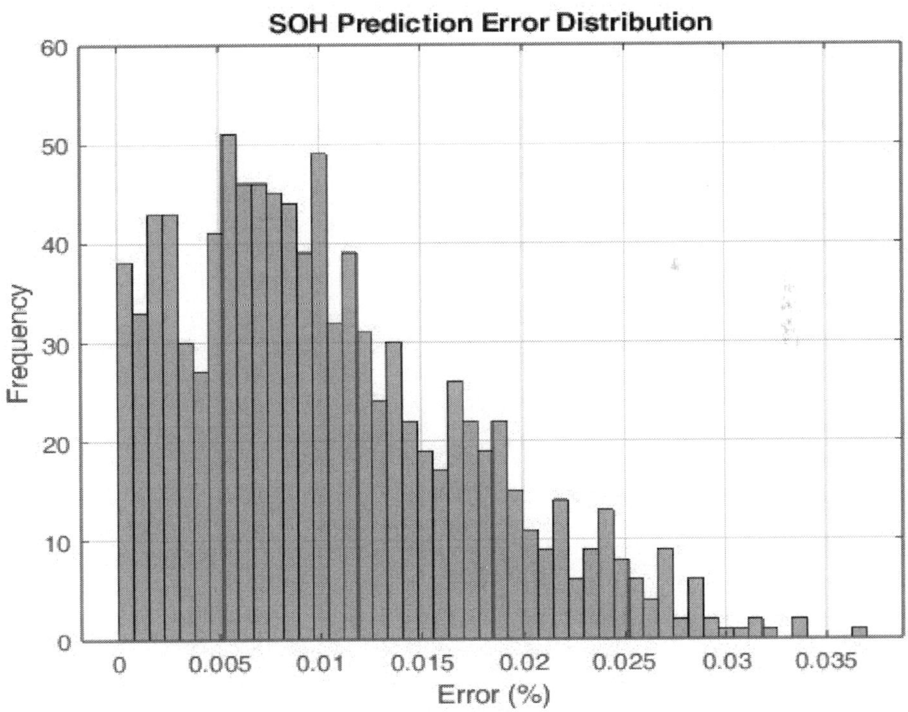

Figure 2.7. *SOH prediction error distribution*

2.8.7. *RUL prediction error analysis*

The error values are predominantly concentrated below 0.3 cycles, with a peak frequency near zero, indicating highly accurate predictions by the model in most cases. A secondary cluster is observed around the 1.0–1.2 cycles range, suggesting the presence of a few outliers that marginally deviate from the model's general performance. Despite this minor deviation, the overall distribution remains sharply skewed toward minimal error, with negligible frequency beyond 1.5 cycles. This tight error margin and limited dispersion signify the model's ability to reliably capture degradation trajectories and forecast battery life with high precision. Accurate RUL estimation is vital for planning maintenance schedules, ensuring operational safety and optimizing battery replacement strategies in EV applications. Figure 2.8 shows RUL prediction error distribution.

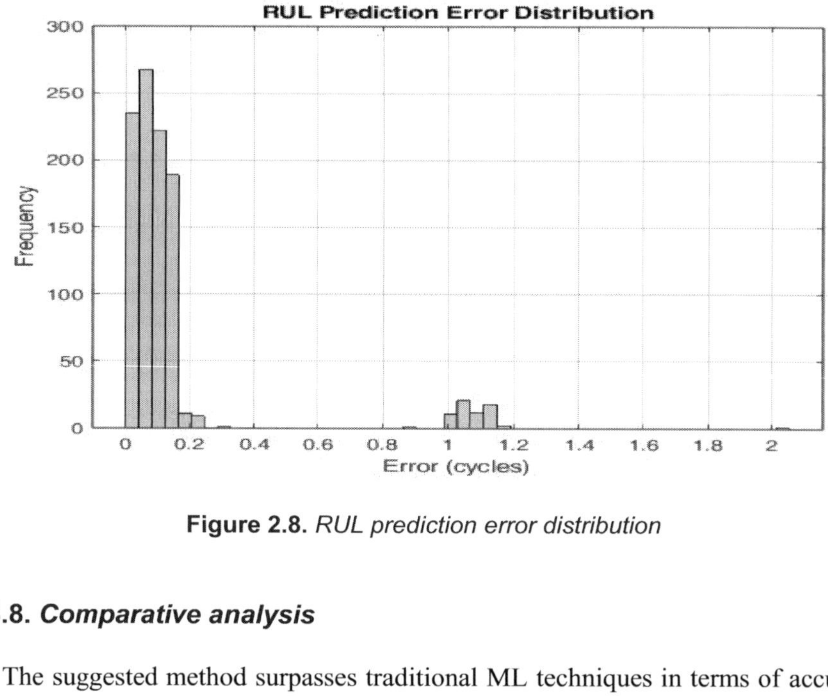

Figure 2.8. *RUL prediction error distribution*

2.8.8. *Comparative analysis*

The suggested method surpasses traditional ML techniques in terms of accuracy and robustness. The use of stacked ensemble models, feature engineering approaches and dataset balance greatly improves prediction reliability. Compared to traditional SOC and SOH estimation models, which often exhibit larger errors due to data imbalance and sensor noise, the proposed approach provides superior generalization capabilities, making it a promising solution for intelligent BMSs. Table 2.1 shows the SOC prediction comparison, and Table 2.2 shows the SOH prediction comparison.

2.8.9. *SOC prediction comparison*

Model	Temp (°C)	SOC MAE (%)	SOC RMSE (%)	Remarks
FCNN	0	0.05846	0.04526	Highest error
LSTM	0	0.04532	0.03148	Moderate performance
GRU	0	0.03911	0.02523	Better than LSTM
LSTM-GRU	0	0.03085	0.02965	Best among DL models at 0°C
Our proposed study	0	0.0181	0.0236	Best overall at 0°C
FCNN	25	0.11573	0.09726	Worst at 25°C
LSTM	25	0.02708	0.02245	Good
GRU	25	0.03444	0.02911	Acceptable
LSTM-GRU	25	0.02066	0.01725	Best among base models at 25°C
Our proposed study	25	0.0208	0.0251	Nearly same MAE, slightly higher RMSE
FCNN	40	0.02592	0.01823	Improved over 25°C
LSTM	40	0.01360	0.01098	Strong performance
GRU	40	0.01218	0.01001	Best SOC at 40°C among base models
LSTM-GRU	40	0.01414	0.01176	Second best
Our proposed study	40	0.0185	0.0247	Slightly higher, but stable across all conditions

Table 2.1. *SOC prediction comparison*

2.8.10. *SOH prediction comparison*

The SOH estimation performance of the suggested model was also compared to the models presented in the base study as in Table 2.2. The LSTM, GRU and LSTM-GRU models were reported to deliver RMSE values of approximately 0.9%, 0.88% and 0.41%, respectively, and MAE values of around 0.75%, 0.72% and 0.26%. In contrast, the proposed hybrid ensemble model demonstrated superior estimation capability with an MAE of only 0.0104% and an RMSE of 0.0125% at 25°C. This reflects a >95% reduction in prediction errors compared to the best-performing DL model in the base paper. This substantial improvement is achieved using interpretable, lightweight models, thereby reducing computational overhead while improving accuracy.

Model	SOH MAE (%)	SOH RMSE (%)	Remarks
FCNN	~0.75	~0.90	Highest error
LSTM	~0.72	~0.88	Slightly better
GRU	~0.70 (est.)	~0.85 (est.)	Estimated, no exact values
LSTM-GRU	0.26	0.41	Best in base paper

Table 2.2. *SOH prediction comparison*

2.8.11. *Fault detection accuracy comparison among classifiers*

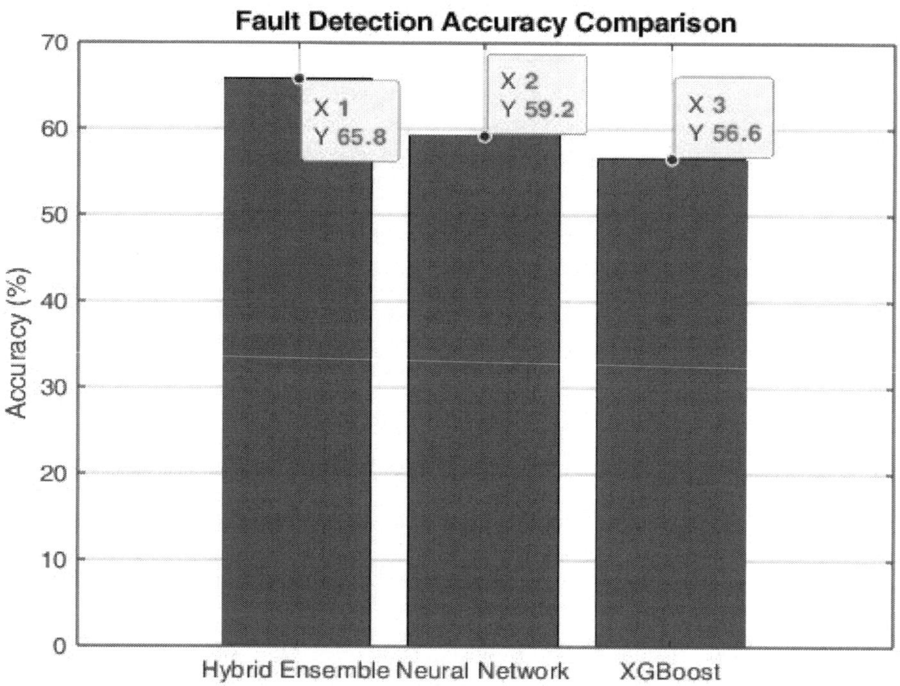

Figure 2.9. *RUL prediction error distribution*

Figure 2.9 presents a comparative analysis of fault detection accuracy achieved using three different ML models: a hybrid ensemble model, a neural network-based model and an XG Boost-style AdaBoost model. The findings clearly show that the hybrid ensemble classifier performed better than the neural network (59.2%) and the XG Boost-style model (56.6%), with an accuracy of 65.8%. The hybrid ensemble model's higher precision can be due to its ability to combine the strengths of many base learners, capturing complicated nonlinear patterns while minimizing overfitting. In contrast, the standalone neural network and XG Boost-style classifiers, while effective, may have disadvantages such as susceptibility to data imbalance and overfitting of noisy features. These findings support the usefulness of the suggested hybrid architecture in defect detection scenarios involving EV battery data. The results reinforce the model's robustness and suggest that ensemble learning, when carefully tuned, can lead to significant performance gains in battery health monitoring applications.

2.9. Conclusion

In this chapter, we propose an advanced BMS for EVs based on a hybrid ML framework that combines stacked ensemble learning, XG Boost and DL techniques to predict accurate SOC, SOH and RUL, as well as fault detection. The methodology used synthetic data generation, innovative feature engineering and ensemble learning techniques to improve forecast accuracy and robustness. The proposed model was evaluated using MAE and RMSE for SOC, SOH and RUL estimation, while fault detection accuracy was SOC prediction: MAE 0.0208%, RMSE 0.0251% SOH prediction: MAE 0.0104%, RMSE 0.0125% RUL prediction: MAE 0.15 cycles and RMSE 0.30 cycles fault detection accuracy: significant improvement using hybrid ensemble model. These results indicate that the proposed hybrid approach effectively captures battery dynamics, improves predictive performance and enhances fault detection capabilities. The integration of data augmentation techniques, feature engineering and ensemble learning has led to a robust and scalable solution for next-generation EV BMSs.

Despite achieving high predictive accuracy and improved fault detection, there are several areas where this research can be extended further: such as real-world validation, online implementaiton for real-time BMS, fault prediction and diagnosis, Battery thermal management system, integration and hybrid energy storage optimization.

Real-world validation: the proposed model has been evaluated using synthetic data. Future research should focus on testing the model with real-world EV battery datasets under various operating conditions. Deep learning integration advanced recurrent neural networks (RNNs), LSTM and transformer models may increase the temporal accuracy of SOC and SOH forecasts.

Online implementation for real-time BMS: implementing the proposed framework in an embedded system or cloud-based infrastructure would enable real-time state estimation and predictive maintenance in EVs.

Fault prediction and diagnosis: instead of fault detection, future models could be trained for fault prediction – anticipating failures before they occur using anomaly detection techniques.

Battery thermal management system (BTMS) integration: extending this work by incorporating battery thermal management models would improve battery lifespan and optimize energy efficiency in EVs.

Hybrid energy storage optimization: the proposed methodology can be extended to optimize hybrid energy storage systems (HESS) that integrate Li-ion batteries with supercapacitors, improving energy efficiency and EV performance. By addressing these challenges, future work can contribute toward a more intelligent, adaptive and self-learning BMS, ensuring longer battery life, higher efficiency and safer EV operations.

2.10. References

Brillianto Apribowo, C.H., Hadi, S.P., Wijaya, F.D., Bambang Setyonegoro, M.I., Sarjiya (2024). Early prediction of battery degradation in grid-scale battery energy storage system using extreme gradient boosting algorithm. *Results in Engineering*, 21, 101709.

Dong, Y., Zhang, G., Li, R. (2024). Comparative study of joint estimation of state of charge (SOC) and state of health (SOH) of lithium-ion batteries based on different tree models. *Journal of Electrical and Electronic Engineering*, 12(2), 23–35. doi: 10.11648/j.jeee.20241202.11.

Jafari, S. and Byun, Y.C. (2022). XGBoost-based remaining useful life estimation model with extended Kalman particle filter for lithium-ion batteries. *Sensors*, 22(23), 9522. doi: 10.3390/s22239522.

Jafari, S., Byun, Y.C., Ko, S. (2023). A novel approach for predicting remaining useful life and capacity fade in lithium-ion batteries using hybrid machine learning. *IEEE Access*, 11, 131950–131963.

Jorkesh, S., Ahmed, R., Habibi, S., Hosseininejad, R., Xu, S. (2025). Battery state of charge and state of health estimation using a new hybrid deep neural network approach. *IEEE Access*, 13, 12566–12580

Meng, J., Cai, L., Yang, S., Li, J., Zhou, F., Peng, J., Song, Z. (2024). An empirical-informed model for the early degradation trajectory prediction of lithium-ion battery. *IEEE Transactions on Energy Conversion*, 39, 2299–2311.

Nkechinyere Njoku, J., Ifeanyi Nwakanma, C., Kim, D.S. (2024). Explainable data-driven digital twins for predicting battery states in electric vehicles. *IEEE Access*, 12, 83480–83501. doi: 10.1109/ACCESS.2024.3413075.

Pradeep, J., Harsha, S.S., Prabhu, R.J., Subramani, K., Mahendran, G., Selvaganapathi, T. (2022). Comparison of SOH prediction ML technique for electric vehicles. In *2022 Sixth International Conference on I-SMAC (IoT in Social, Mobile, Analytics and Cloud) (I-SMAC)*, Dharan, 510–515. IEEE Xplore. doi: 10.1109/I-SMAC55078.2022.9987283.

Ren, J., Ma, J., Wang, H., Yu, T., Wang, K. (2025). A comprehensive review on research methods for lithium-ion battery of state of health estimation and end of life prediction: Methods, properties, and prospects. *Protection and Control of Modern Power Systems*, 10, 146–165. doi: 10.23919/PCMP.2024.000211.

Surya, S., Saldanha, C.C., Williamson, S. (2021). Novel technique for estimation of cell parameters using MATLAB/Simulink. *Electronics*, 11, 117.

Swarnkar, R., Harikrishnan, R., Thakur, P., Singh, G. (2023). Electric vehicle lithium-ion battery ageing analysis under dynamic condition: A machine learning approach. *SAIEE Africa Research Journal*, 114(1), 4–13. doi: 10.23919/SAIEE.2023.9962788.

Wei, L., Xu, H., Sun, Y., Diao, Q., Tan, X., Fan, Y., Liu, H. (2025). State of health estimation of lithium-ion batteries based on relaxation voltage reconstruction and AM-LSTM method. *IEEE Transactions on Transportation Electrification*, 11, 7261–7273.

Yang, G., Ridgeway, C., Miller, A., Sarkar, A. (2024). Comprehensive assessment of artificial intelligence tools for driver monitoring and analyzing safety critical events in vehicles. *Sensors*, 24, 2478. doi: 10.3390/s24082478.

Yang, W., Min, F., Xie, J., Yang, H. (2025). Ultra-early prediction of lithium-ion battery cycle life based on visualized single-cycle data. *IEEE Transactions on Power Electronics*, 40, 7342–7353. doi: 10.1109/TPEL.2025.3531791.

Zhang, H., Niyato, D., Zhang, W., Zhao, C., Du, H., Jamalipour, A., Sun, S., Pei, Y. (2025). The role of generative artificial intelligence in Internet of Electric vehicles. *IEEE Internet of Things Journal*, 12, 6208–6232. doi: 10.1109/JIOT.2024.3511961.

3

Operation and Control of an Autonomous Hybrid Generation System Based on Wind, Solar, Biogas and Battery Bank

This study introduces a novel AHRGS (autonomous hybrid renewable generation system) for rural power distribution, integrating solar, wind, biogas and battery storage. A doubly fed induction generator (DFIG) is used to transform wind energy, with PV panels as the primary source, supported by a biogas generator (BG) and battery bank (BB). A converter Output Current Based Maximum Power Point Tracking (COCB-MPPT) algorithm exploits the dynamics between the current output of the converter and duty ratio to routinely fine-tune step fluctuations, thus maximizing power delivered from the PV array. This approach determines the peak electrical power of the PV array merely by monitoring the battery charging current delivered to the battery via the boost converter. In the proposed approach, a VFCL (voltage and frequency closed loop) controller combined with a harmonic filter is employed as a voltage controller to reduce the THD percentage. Development of AHRGS based on a DFIG architecture that utilizes a single-inverter battery configuration eliminates the need for conventional dual-converter systems. The VFCL system does not require rotor position detection, flux estimation or coordinate transformations, making it cost-effective and easier to implement. The AHRGS topology is simulated in MATLAB and experimentally validated under various operating conditions. Test results presented indicate the VFCL control scheme effectively ensures steady stator voltage at 415 V and the frequency at 50 Hz under all conditions. Further, measured THD values remain below 3% for voltage and below 6.5% for current. Power discrepancies between the solar source and the DFIG rotor are balanced in real-time by charging or discharging the batteries, ensuring a stable and uninterrupted power supply to isolated loads. The study includes active and reactive components of power supply using the DSTATCOM, supplying reactive power from the capacitor bank, and the wind power output becomes zero. The outcomes and associated waveforms are discussed in detail. Furthermore, the results confirm that the VFCL controller, along with a harmonic filter, is an impactful approach to mitigate voltage uncertainty.

Chapter written by Nindra SEKHAR, Kumaresan N, Venkatesh BODDAPATI, Ravulakari KALYAN, Venkata Madhavaram T and Nethravathi S.

3.1. Introduction

Increasing the proportion of renewable energy in the global energy mix by 2030 demands a multifaceted approach involving robust policies, cutting-edge technological innovations, significant investments and heightened public awareness (Hoseinzadeh et al. 2022; Hassan et al. 2024). Advances in renewable energy technologies, for example, next-generation solar panels, wind turbines and energy storage solutions, promise to drive down costs and boost efficiency (Moosavian et al. 2024). However, the full potential of these advancements can only be realized through the strategic integration and complementation of both conventional and renewable energy sources.

The integration of renewable energy systems is vital for India, where a considerable segment of the population exists in geologically isolated and rural areas. These regions often grapple with limited access to electricity, inadequate electrification infrastructure and frequent power outages, which hinder socioeconomic development and deepen inequalities. However, these areas hold important potential for renewable energy generation, with plentiful natural resources such as sunlight, wind and biogas providing a sustainable way to address energy access challenges. Deploying renewable technologies like solar photovoltaic (PV) systems, wind turbines and biogas solutions locally can reduce reliance on centralized grid infrastructure while empowering rural communities with reliable electricity for homes, schools, healthcare facilities and industries (Hassan et al. 2024).

Despite their promise, renewable energy sources are inherently variable, as their output depends on ecological, seasonal and daily weather patterns. This variability poses challenges in meeting the consistent energy demands of remote areas, especially in regions where the main-grid electricity supply is unreliable or unavailable (Javaid et al. 2024). To overcome these limitations, integrated energy solutions are essential. Hybrid systems that combine multiple renewable sources, supported by energy storage technologies such as batteries or pumped hydro storage, can help stabilize energy supply by compensating for fluctuations. Additionally, microgrids and decentralized energy systems can tailor energy generation and distribution to local needs, enabling a sustainable and resilient energy framework (Zhao et al. 2024). By adopting such integrated approaches, India can effectively electrify remote areas, advance its energy security and climate goals, and foster inclusive growth and rural development.

Many developed and undeveloped populations in isolated areas depend on traditional BGs for their power supply. These generators produce electricity using organic materials such as wood, agricultural residues and animal waste. This method of power generation is invaluable in remote locations where access to conventional grid electricity is limited or nonexistent (Ullah et al. 2021).

Biogas energy, when implemented on a large scale, is a reliable and renewable energy source that operates independently of climatic conditions, making it a complementary energy source to intermittent renewables like solar and wind. By combining renewable sources with conventional energy sources such as biogas, a more reliable and continuous power supply can be achieved (Al-Ammar et al. 2022).

Integrating biogas energy with renewable sources significantly reduces reliance on fossil fuels. When managed sustainably, ensuring that the biogas consumption rate does not exceed its regeneration capacity, it causes minimal environmental harm. Many investigators have explored the prospect of biogas energy for optimizing energy systems, highlighting its role in creating more robust and sustainable electric power infrastructures (Wang et al. 2022).

The effective integration of conventional (grid or biogas-derived) and renewable (solar, wind, etc.) energy sources relies heavily on the availability of reliable energy storage systems. Recent advancements in high-capacity batteries and sophisticated battery management systems are paving the way for more efficient integration of multiple energy sources (Chauhan et al. 2013; Silva et al. 2020). By efficiently storing excess energy produced by renewables throughout maximum generation intervals, these storage solutions allow power grids to draw on these reserves when renewable output declines, effectively mitigating the intermittency challenges associated with renewable energy. Recent studies indicate that the adoption of large-scale batteries can significantly reduce renewable energy infrastructure costs while meeting a substantial portion of the regional power demand with sustainable energy sources (Shayan et al. 2023; Ali et al. 2025).

Wind power generation is instrumental in endorsing innovation and research through advancements in the design of wind turbines, materials and improving response. Exploratory and applied research aim to refine turbine structures, improve energy absorption efficiency and integrate wind energy systems coupled with storage and smart grid infrastructure (Yu et al. 2022). The indicated innovations enhance the performance and cost-effectiveness of wind energy while driving technological progress across other industries.

Generating electrical energy through solar PV panels is essential for cleaning up the energy system and advancing ecological development. Solar energy is widely regarded as an encouraging elucidation for minimizing greenhouse gas emissions and confronting climate change (Yuan et al. 2024). Its broad adoption can play a pivotal role in achieving these critical objectives. By replacing traditional energy sources like coal and natural gas with PV solar energy, it is possible to substantially lower CO_2 emissions, facilitating cleaner and more climate-conscious energy utilization.

Electricity generation through biogas offers a promising and sustainable approach to reducing CO_2 emissions resulting from energy sector activities. Harnessing the full biogas potential requires sustainable practices, including responsible feedstock sourcing and the implementation of advanced technologies to minimize environmental impacts. By encouraging investments in biogas energy infrastructure and research, governments and industries can support a transition to greener energy systems. In doing so, biogas plays a vital support in fulfilling global climate goals and fostering a tough, low-carbon future (Pérez et al. 2025).

The amalgamation of solar and wind energy with biogas and battery storage systems presents a robust strategy to tackle the challenges posed by renewable energy intermittency in the rural areas of India (Diab et al. 2019; Fernández-Muñoz et al. 2021; Sekhar et al. 2022). This synergistic approach not only ensures a reliable and high-quality power supply by balancing the fluctuations in solar and wind energy with the stability of biogas and the adaptability of battery storage but also delivers both environmental and economic advantages, positioning it as a compelling strategy for sustainable energy development (Fernández-Muñoz et al. 2023; Xu et al. 2023). The proposed off-grid hybrid generation system employs a single converter at the rotor side of DFIG for effective control (Diab et al. 2019; Hassan et al. 2023).

Currently, DFIGs are suggested for wind energy conversion systems owing to the abbreviated grading of back-to-back converters and functioning across both super-synchronous and sub-synchronous rotor speeds. DFIGs are more and more engaged in electric power generation with wind energy sources, both for stand-alone and for grid implementation (Kant et al. 2017; Tiwari et al. 2019). Considering the rising reputation for isolated systems set-up of WECS, a modest control method is anticipated in the proposed article for powering isolated loads through DFIGs. The topology for the proposed hybrid system is selected in such a way that it requires single bi-directional converter, that is, an SPWM inverter for integrating DC sources (solar and battery) with an AC supply (Vijayakumar et al. 2012, 2013b; Sharma et al. 2017).

This article explores a novel AHRGS in an off-grid setting. Previous studies on hybrid BM-PV-wind-biogas systems have primarily intensive on energy supervision, optimal sizing and economic investigation. In contrast, a study in Sekhar et al. (2022) proposed voltage control for such microgrids using general parameters from past research but did not account for the power output of individual renewable energy sources or the current THD.

In this study, the model from Sekhar et al. (2022) is adapted while incorporating power output and THD analysis for each renewable energy source. Additionally, a VFCL controller is implemented as the voltage control mechanism due to its advantages, specifically simplicity, fast transient response and superior voltage stability.

The principal purpose of this research is to study a novel AHRGS configuration and assess voltage stability using a VFCL controller through evaluating output power, voltage of the system and current THD levels. The key role of this study is demonstrating the successful mitigation of voltage uncertainty in the anticipated microgrid system through the combined use of a VFCL controller and a power harmonic filter.

In this context, this research presents a novel AHRGS premeditated to assist the electrification of limited rural areas, with a particular focus on rural regions. The purpose of this chapter is to examine the operation of the AHRGS for stand-alone applications. The specific aims of this work include:

1) Integrating a variety of renewable power sources, including solar, wind, biogas and battery storage.

2) Implementing the regulation strategy for the DFIG on an FPGA controller (Altium Nanoboard 3000) using hardware description language (HDL).

3) Conducting extensive experimentation to corroborate the integration of the wind-solar hybrid generation system.

4) The VFCL control approach preserves the stator voltage at 415 V and the frequency at 50 Hz, each within a ±1% tolerance, across all operating conditions. Experimental results show that the measured THD values remain below 3% for voltage and below 6.5% for current.

Section 3.2 discusses power generation and modeling aspects of the autonomous hybrid renewable generation system, including wind, solar PV, battery, and biogas components. Section 3.3 presents the comprehensive operation of the proposed AHRGS, while section 3.4 details the control strategy. Simulation results across different scenarios are presented in section 3.5, and section 3.6 describes the experimental outcomes involving solar PV, integrated solar PV with DFIG, and the overall hybrid system. Finally, section 3.7 offers a comprehensive discussion pertaining to findings.

3.2. Autonomous hybrid renewable generation system

This schematic showcases a multi-source renewable energy system designed for reliable electricity generation. Wind turbines convert wind energy into electricity, while PV panels capture the sun's rays and convert them into electrical currents. A biogas engine utilizes organic material like food waste or animal waste to create electricity through a squirrel cage induction generator (SCIG). In the proposed system, as depicted in Figure 3.1, DFIG stator windings are directly linked to the PCC, while the rotor windings are connected to a SPWM inverter. This inverter, in turn, is connected to the PV panels. Maximum power is extracted using a boost converter, which is in turn connected to the BB. By combining these diverse renewable sources with battery storage, the system aims to provide a dependable and sustainable electricity supply with reduced reliance on fossil fuels.

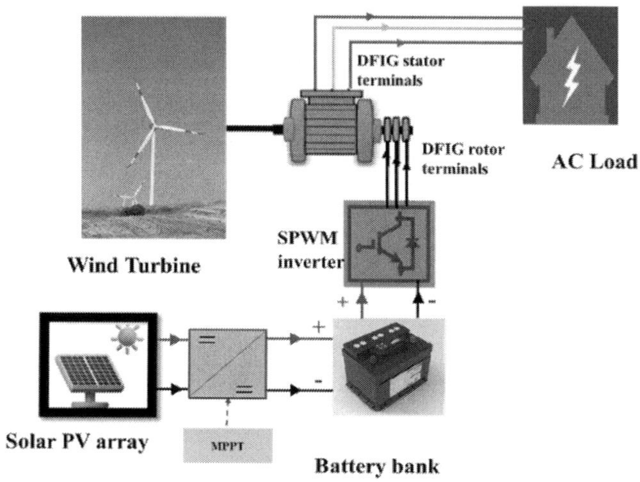

Figure 3.1. *AHRGS with load. For a color version of this figure, see www.iste.co.uk/sundaram/energysystems.zip*

3.2.1. *Wind energy system*

A wind turbine comprises rotor blades connected to a generator via a gearbox, which increases rotor speed for power generation. A control system regulates output. Power depends on wind speed, rotor speed, blade pitch, tilt and turbine geometry, forming a model vital for control and analysis. At a steady rate of acceleration a, kinetic energy Ek of an object with mass (m) and velocity (v) amounts to the work performed (W) in shifting the object from rest to a certain distance (d) experiencing force (F), that is, Ek = W Fd as stated by Newton's second law of motion (Manyonge et al. 2012).

$$F = ma \qquad [3.1]$$

Hence, the kinetic energy is given as:

$$E_k = mad \quad [3.2]$$

Using principles of solid motion kinematics $v^2 = u^2 + 2ad$, where u is the object's velocity at the beginning. Therefore, $a = \frac{v^2-u^2}{2d}$. Considering zero initial velocity, then $= \frac{v^2}{2d}$. Therefore, from equation [3.2]:

$$E_k = \frac{1}{2}mv^2 \quad [3.3]$$

In this work, it is assumed that air density remains approximately constant despite changes in altitude or temperature. Thus, by applying the kinetic energy as expressed in equation [3.3], the kinetic energy (in joules) of air with mass m moving at wind velocity v_{wind} is calculated using this equation. The wind power P_{wind} is then defined as the rate of variation in kinetic energy, that is:

$$P_{wind} = \frac{dE_k}{dt} = \frac{1}{2}\frac{dm}{dt}v_{wind}^2 \quad [3.4]$$

However, to express power in terms of wind velocity and air density, the mass flow rate $\frac{dm}{dt}$ is expressed as $\frac{dm}{dt} = \rho A v_{wind}$. Here, A represents the cross-sectional area of wind flow and ρ denotes the air density. Using this expression, equation [3.4] can be rewritten as:

$$P_{wind} = \frac{1}{2}\rho A v_{wind}^3 \quad [3.5]$$

$$P_m = \frac{1}{2}\rho A v_{up}^3 C_p \quad [3.6]$$

where:

$$C_p = \frac{\left(1+\frac{v_{down}}{v_{up}}\right)\left(1-\left(\frac{v_{down}}{v_{up}}\right)^2\right)}{2} \quad [3.7]$$

In equation [3.7], C_p denotes the portion of upstream wind power harnessed by the rotor blades and is known as the Betz limit.

3.2.1.1. *DFIG configuration*

Wind turbines and wind power generators are employed to capture wind energy, complemented by converters for the conversion of mechanical energy into electricity, as depicted in Figure 3.2. The DFIG stands out as a widely adopted wind power generation system within Wind Energy Conversion Systems (WECS), owing to its notable advantages. These include its high energy conversion efficiency across diverse wind speeds, capability to improve power quality, relatively low converter cost and minimized power losses compared to systems utilizing fully fed synchronous generators.

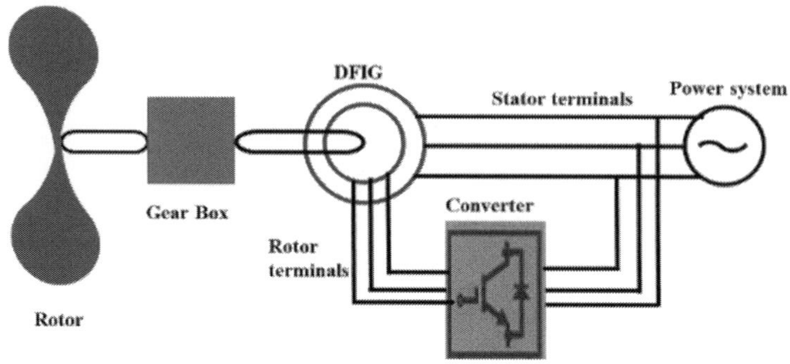

Figure 3.2. *Schematic diagram of the DFIG system*

3.2.2. *Diode-based PV cell model*

The one-diode model for the solar cell model equivalent circuit for demonstrating PV modules is well known, and it is depicted in Figure 3.3. The relationship between voltage and current at a particular irradiation and temperature is given as:

$$I_{input} = I_{PH} - I_D - I_P \quad [3.8]$$

$$I_{input} = I_{PH} - I_{rs}\left[e^{\frac{V_{input}+R_S I_{PV}}{N_S b V_T}} - 1\right] - \frac{V_{input}+R_S I_{PV}}{R_{SH}} \quad [3.9]$$

Here, V_{input} represents the voltage and current output from the solar PV module, respectively. IPH denotes the photocurrent generated by incident light, while I_{rs} corresponds to the reverse saturation current of the solar cell. In an ideal PV module, the series resistance is presumed to be zero, and the parallel resistance is presumed to be infinite. Then equation [3.9] is modified as:

$$I_{input} = I_{PH} - I_{rs}\left[e^{\frac{V_{input}}{N_S b V_T}} - 1\right] \quad [3.10]$$

$$I_{PH} = [I_{sh\,STC} + k_{ish}(T - T_{STC})]\frac{G}{G_{STC}} \quad [3.11]$$

By examining equations [3.10] and [3.11], it becomes evident that the voltage of the PV array exhibits a nonlinear relationship with irradiation. In contrast, the current of the PV array demonstrates a linear relationship with irradiation.

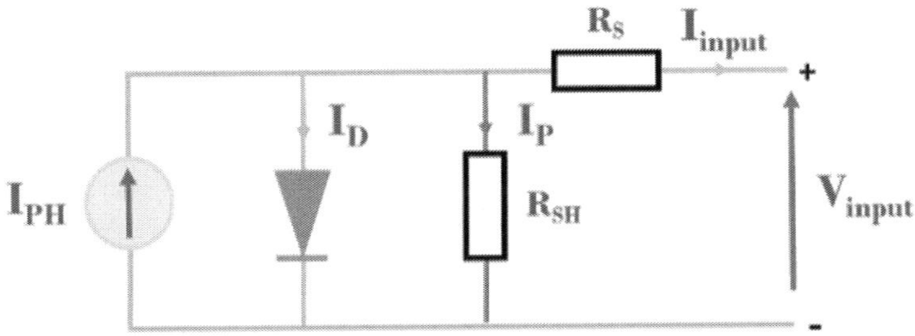

Figure 3.3. *The single diode model for a photovoltaic (PV) cell. For a color version of this figure, see www.iste.co.uk/sundaram/energysystems.zip*

Equations [3.10] and [3.11] show that PV array voltage is nonlinear with irradiation, while PV array current is linear with irradiation, and it is clearly outlined in Sekhar et al. (2025).

3.2.3. *Battery model*

The circuit model, as shown in Figure 3.4, epitomizes the battery by assembling basic electrical components into a simplified circuit network. It approximates both the internal electrochemical processes and the peripheral dynamic performance of the battery as anticipated in equations [3.12] and [3.13].

As stated by Kirchhoff's law:

$$V_0 = I_L R_0 \qquad [3.12]$$

where

$$I_L = \frac{V_1}{R_1} + C_1 \frac{dV_1}{dt}$$

$$I_L = \frac{V_2}{R_2} + C_2 \frac{dV_2}{dt}$$

$$V_L = V_{BOC} - V_0 - V_1 - V_2 \qquad [3.13]$$

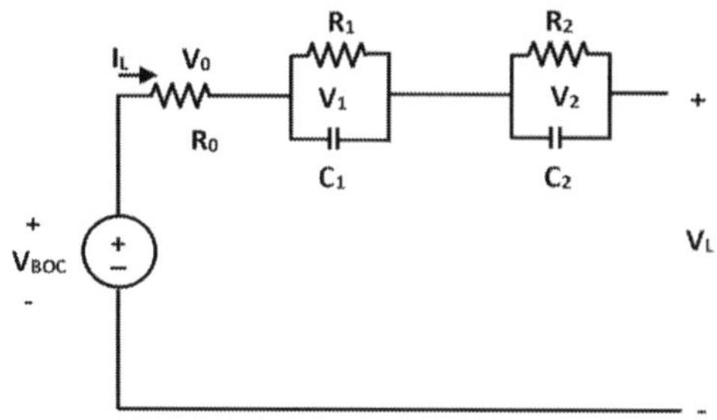

Figure 3.4. *Battery equivalent circuit model*

3.2.4. *Biogas energy*

Biogas is a type of biofuel that forms naturally from the anaerobic decomposition of biological waste. While organic substances, including biodegradable waste from food and animals, disintegrate in an oxygen-deprived environment, they discharge a mixture of gases, primarily methane and carbon dioxide. As a result of the rich concentration of methane in biogas is flammable, and consequently yields a deep blue flame, and is suitable for use as a fuel source. The Biogas power plant is depicted in Figure 3.5.

Mathematical models, as anticipated from equation [3.16], offer a cost-effective and efficient alternative to full-scale experiments for conducting simulation studies (Marzhan et al. 2020). The commonly used models describe the reaction kinetics of methane fermentation and the related kinetic factors. Diverse facsimiles define the development and enlargement of the micropopulation, the components' degeneration and the groundwork of yields in an anaerobic bioreactor.

– S = feed substrate concentration (g/L).

– Sh = hydrolyzed product concentration.

– Si = S1, S2, S3, etc., are the concentrations (g/L) of different raw materials.

– P = aggregate biogas output (g/L).

– k = kinetic rate constant for hydrolysis.

– ki = kinetic rate constant for hydrolysis for precise to Si categories.

– T = total number of raw material types.

– $f_H(S_i), f_M(S_i)$ = hydrolysis reactions, reticence and microbial fermentation development by fatty acids.

– A_h = substrate with elevated solid content intended for processing in a high-rate anaerobic digester.

– A_m = substrate with elevated solid content intended for processing in a medium-rate anaerobic digester.

The following modeling method is considered for biofuel production from organic municipal waste:

$$\frac{\partial s_i}{dt} = k_i.W_i.f_H(S_i), i = 1 \dots \dots, T, S(0) = S_0 \quad [3.14]$$

where:

$$f_H(S_i) = \left(1 + \left(\frac{S_h}{A_h}\right)^{A_h}\right)^{-1}, f_M(S_i) = \left(1 + \left(\frac{S_h}{A_m}\right)^{A_m}\right)^{-1}$$

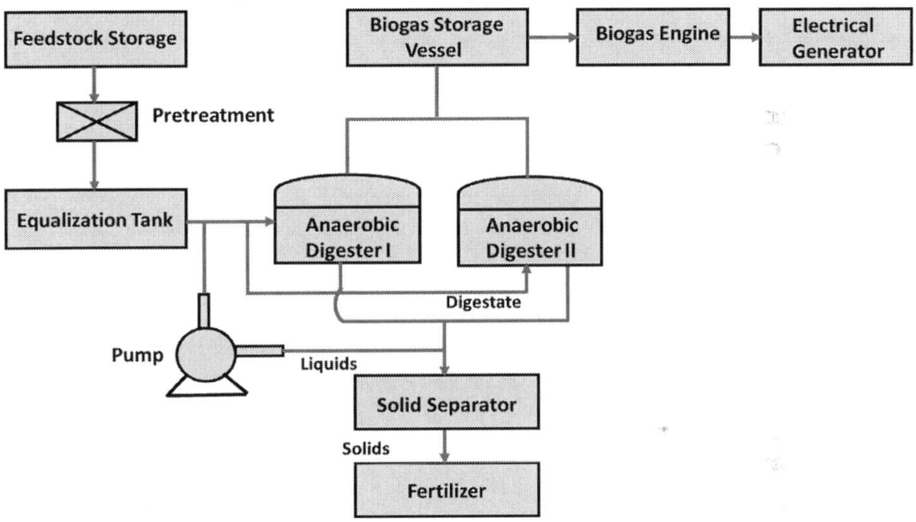

Figure 3.5. *Block diagram of biogas power generation*

3.3. Comprehensive operation of the proposed AHRGS

The hybrid renewable power generation system combines diverse renewable energy sources with conventional energy generation and storage technologies to ensure a reliable and sustainable power supply. The system comprises a DFIG for wind energy extraction, a SCIG running on a biogas engine, PV panels for solar energy extraction and a BB for energy storage.

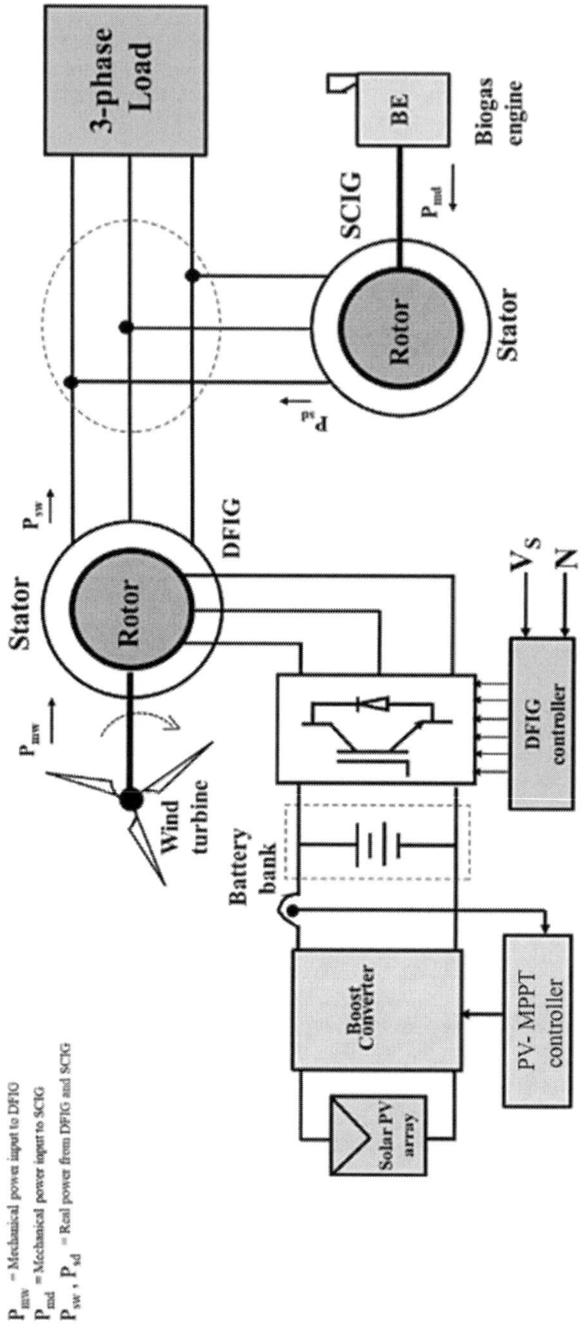

Figure 3.6. *Wind, solar and biogas hybrid renewable system. For a color version of this figure, see www.iste.co.uk/sundaram/energysystems.zip*

Both generators are interconnected with isolated AC loads and stator terminals coupled to form the PCC. The solar PV panels are linked to a BB. This integration allows for the efficient extraction and storage of solar energy. The PV-battery system is interfaced with the DFIG rotor terminals through an SPWM inverter, enhancing the system's overall efficiency and stability. SCIG is directly connected to the PCC to supply isolated AC loads. The complete schematic model of the anticipated hybrid power generation system is depicted in Figure 3.6. The DFIG and SCIG have a rated capacity of 3 HP, indicating their capability to generate power efficiently within specified parameters. The hybrid nature of the system allows for optimal utilization of available energy resources while minimizing environmental impact and operational costs.

3.4. AHRGS control strategy

Each renewable energy source in this system employs a specific control strategy, except for biogas generation, which maintains stable voltage and current without additional voltage control due to its predictable input. Table 3.1 summarizes the control strategies for each renewable energy design in this system. The THD value for AC voltage and current in the proposed AHRGS is calculated through the method shown in Figure 3.7.

RE source	Component	Control strategy
Solar PV system	Boost converter	Single sensor MPPT
Solar PV-Wind	SPWM inverter	VFCL control
Solar PV-Wind-Biogas	SPWM inverter	VFCL control

Table 3.1. *AHRGS control method*

3.4.1. *VFCL control strategy*

Preserving persistent voltage and frequency at PCC is crucial to ensure a stable power supply. The anticipated hybrid power generation scheme is attained via suitably adjusting the DFIG rotor voltage and frequency. The VFCL controller is intended to maintain the PCC voltage at 415 V and 50 Hz. As presented in Figure 3.8, the controller arrangement has two loops: one for voltage regulation, whereas the other serves for frequency control. The voltage regulation loop measures the DFIG stator voltage RMS value and computes the error, Verr = Vref – Vrms. This error is processed in the controller to determine the essential modulation index (MI). The DFIG rotor frequency fr is derived from the measured rotor speed.

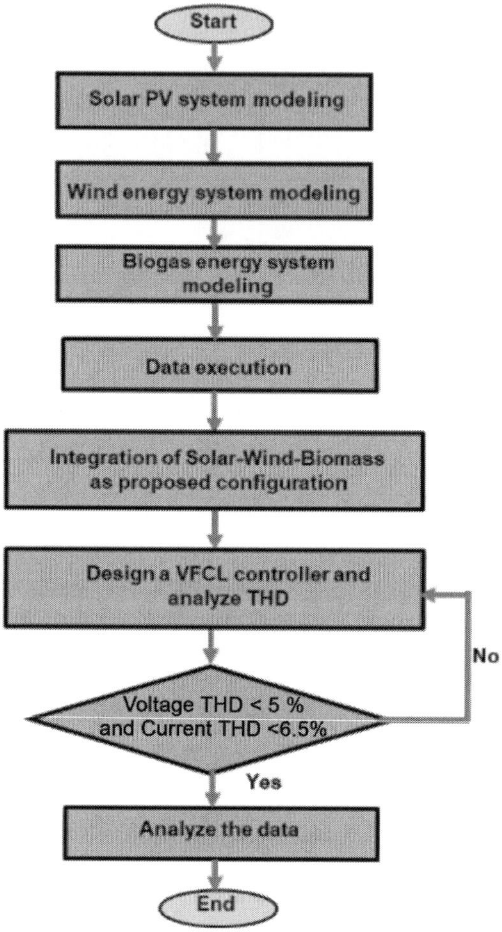

Figure 3.7. *AHRGS methodology*

A unit sinusoidal wave of frequency fr is produced and scaled by MI to produce the reference. This waveform is equated with a carrier wave to breed SPWM pulses, which drive the inverter supplying the rotor. The same is instigated on an FPGA controller using HDL. Unlike the DFIG, the BG is unswervingly coupled to the PCC, devoid of a power electronic regulator. To function as a generator, the SCIG must be rotated at a speed higher than its synchronous speed. As determined by the rotor speed, the SCIG will provide the real power at the PCC and take reactive power for its function. Figure 3.8 illustrates the VFCL controller developed explicitly to preserve constant voltage and frequency at the PCC. This controller continuously monitors the voltage and frequency levels at the PCC and adjusts the DFIG rotor voltage and frequency accordingly to confirm the stability and reliability of the power supply (Sekhar et al. 2019, 2022; Vijayakumar et al. 2013a; Sam et al. 2017; Puchalapalli et al. 2020).

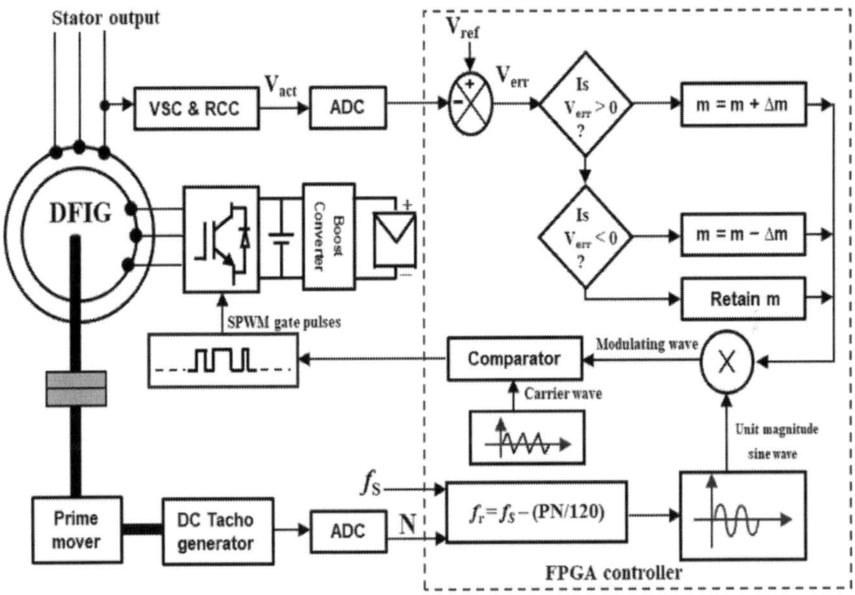

Figure 3.8. *Control strategy for DFIG voltage and frequency*

3.4.2. *Single sensor solar PV-MPPT method*

The current–voltage (I–V) behavior of a PV array is mainly influenced by solar irradiance and temperature. These environmental factors cause shifts in the array's performance, leading to a distinct MPP under each set of conditions. To evaluate the efficacy of the projected COCB-MPPT algorithm, a PV array is coupled to a BB via a DC–DC boost converter. The battery functions as a load, helping to stabilize the system voltage. The converter manages both the voltage and current from the PV array, thereby regulating the power delivered. The COCB-MPPT algorithm dynamically adjusts the converter's duty cycle to confirm operation at the MPP as delineated in Figure 3.9 and explained in Sekhar et al. (2025). Hardware details follow in the next section. Power flows from the PV array to the battery through a boost converter with efficiency, η, operating at 105 V input and 220 V output. For

CCM, calculated values are 54 μF (capacitor) and 5.4 mH (inductor). The setup uses a 470 μF/450V capacitor and a 10 mH/20A inductor to ensure reliable continuous current mode (CCM) operation.

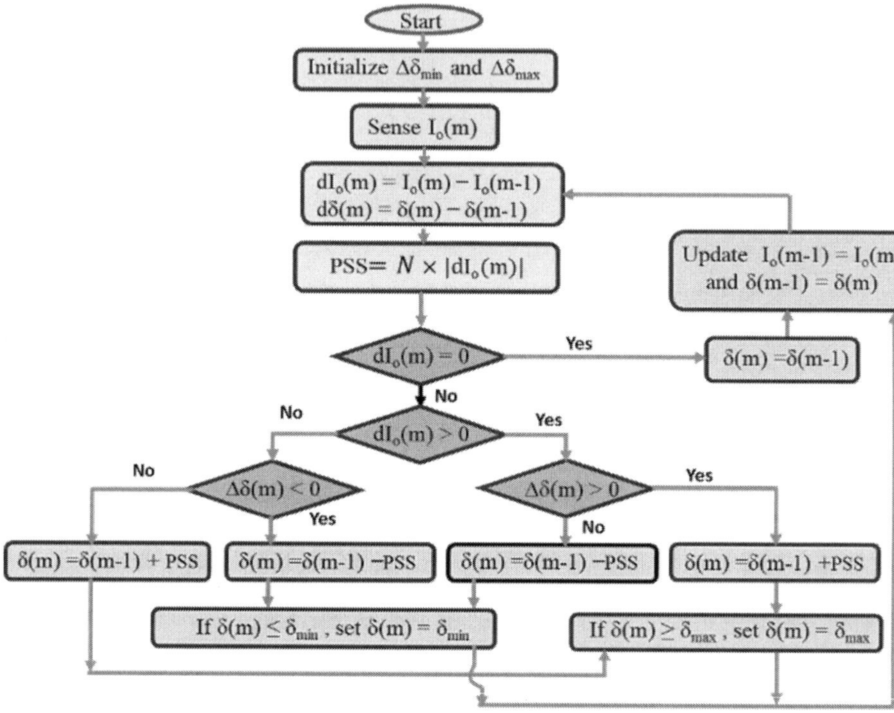

Figure 3.9. *Flow chart for PV-MPPT system. For a color version of this figure, see www.iste.co.uk/sundaram/energysystems.zip*

3.5. Simulation results for the proposed hybrid system

The simulation of the operation and control of a proposed hybrid power generation system is conducted within the MATLAB/Simulink environment. The necessary parameters for conducting the simulation of the anticipated AHGS are outlined in Table 3.2. The simulation results illustrate the performance of supplying three-phase AC loads under various operating conditions. The solar PV panels play a supportive role by providing power to the DFIG rotor. Furthermore, they charge the BB. The BB stabilizes the DC link voltage of the DFIG rotor-side inverter and also assists the PV array in delivering the necessary DC link power. The scenarios considered are as follows: (1) Biogas engine speed variation, (2) wind generator

(DFIG) speed variation and (3) load variation. The voltage and frequency at PCC are constant irrespective of irradiation variation in all circumstances.

Component	Ratings
DFIG	3 HP, Stator: 415 V and 50 Hz;
SCIG	3 HP, Stator: 415 V and 50 Hz;
PV panel	P_{max} = 210.16 W, V_{oc} = 35.7 V, I_{sc} = 7.85 A, V_{mp} = 28.4 V, I_{mp} = 7.4 A
PV array	Ten panels (arranged in series) P_{max} = 2.1 kW, V_{oc} = 357 V, I_{sc} = 7.85 A
Battery bank	Nominal voltage of 440 V and 100 Ah

Table 3.2. *Simulation parameters*

3.5.1. *SCIG speed variation*

In this mode, the synchronous speed of the SCIG is set above the synchronous speed, which is 1,500 rpm, while the desired speed of the DFIG is maintained at 1,500 rpm. The shaft of the biogas engine is mechanically coupled to the shaft of the SCIG, driving the rotor above synchronous speed to initiate and sustain power generation. The reactive power required for SCIG operation is supplied by a DFIG. Under typical grid-connected operation, the SCIG functions as a generator within a limited slip range (up to 6%), corresponding to a speed range of 1,500–1,590 rpm. This operating behavior is taken into account in the proposed off-grid system for selecting the appropriate speed range. The solar intensity set value fluctuates continuously between 300 and 1,000 W/m^2. The load considered in this mode is 2 kW with UPF. The COCB MPPT algorithm is employed to trace the peak power from the PV panels.

The results of this mode are carefully analyzed and elaborated. Figure 3.10(a) depicts the variation of the duty ratio, estimated by the PV-MPPT algorithm, which corresponds to the peak power of the PV panels for different irradiations. The PV panels' output voltage, VPV, remains fixed, facilitated by the stable battery terminal voltage, VBATT, as evident from Figure 3.10(b).

Other parameters from the solar PV system, such as the PV panels' current, IPV, boost converter output current, IBOOST, and PV array power, PPV, are illustrated in Figures 3.10(b)–(d), respectively. It has been noted that the power generated by the solar PV array is utilized to charge the BB whenever it meets the demand for power supply to the DFIG rotor.

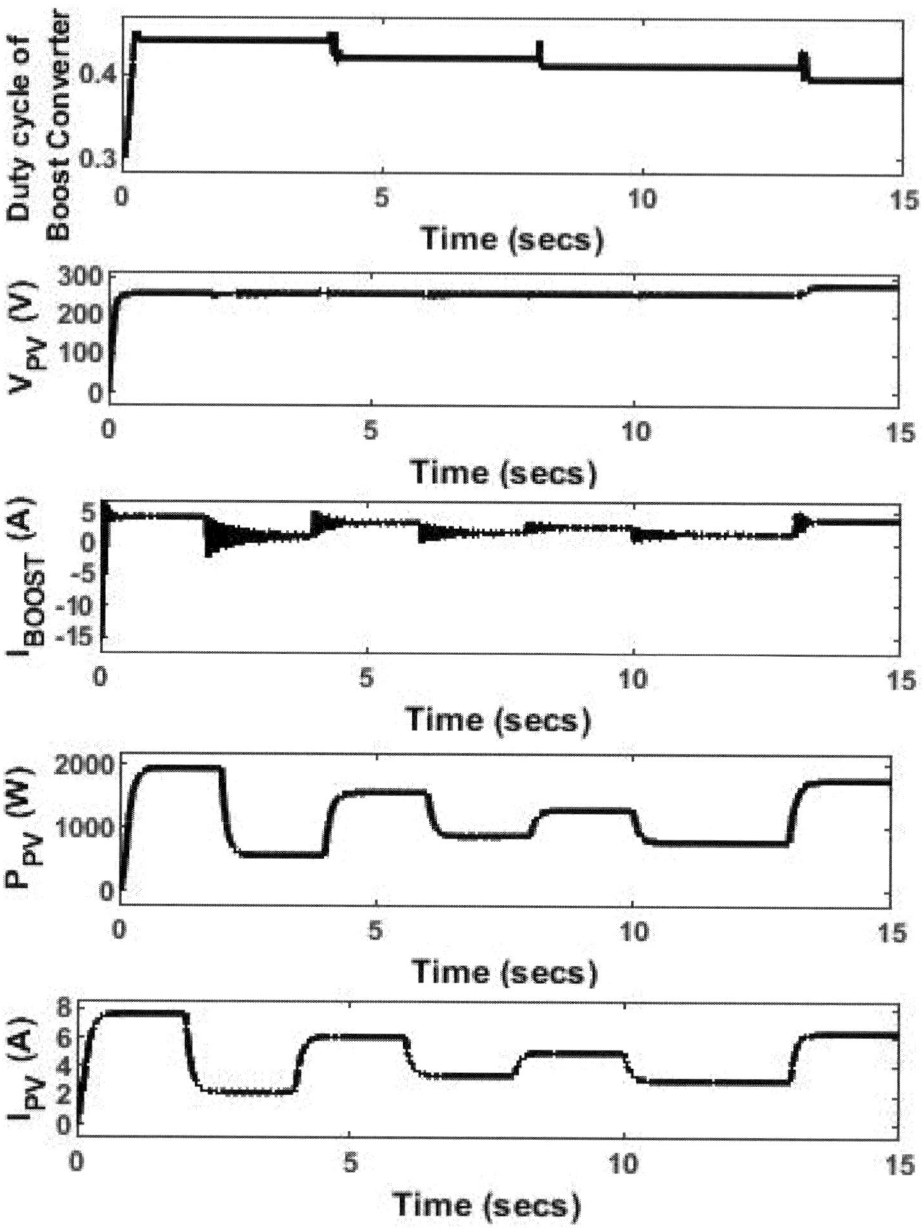

Figure 3.10. *Waveforms of a PV-Boost converter system for different irradiation levels*

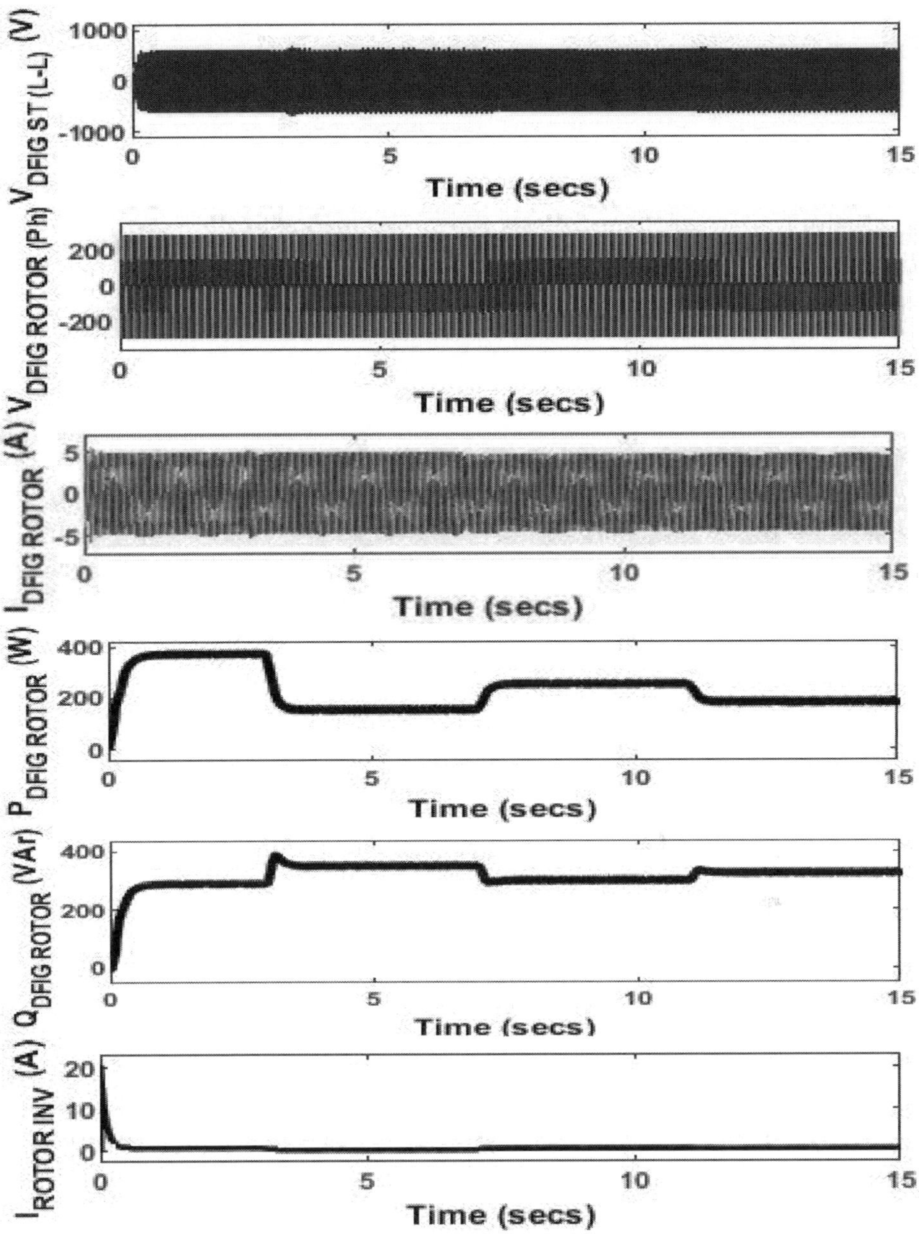

Figure 3.11. *Waveforms at DFIG stator and rotor terminals. For a color version of this figure, see www.iste.co.uk/sundaram/energysystems.zip*

The effective functioning of both the DFIG controller and PV-MPPT controller is demonstrated by the results obtained. The DFIG controller is designed to maintain a constant voltage of 415 V at the PCC, regardless of variations in SCIG speed and under a fixed load condition. Despite changes in the SCIG speed (NSCIG) and reactive power support required for the SCIG, the DFIG closed-loop controller ensures the magnitude of VDFIG_ST (stator voltage of DFIG) is preserved. This highlights the robustness and efficacy of the DFIG control system in maintaining grid voltage stability and supporting the SCIG operation under varying conditions.

The expert functioning of the controller, considering all the mentioned aspects for a load power (PLOAD) of 2 kW at UPF, DFIG speed (NDFIG) of 1,300 rpm and varying SCIG speed (NSCIG), can be observed in Figure 3.11(a). The value of VDFIG_ST is maintained at 415 V, showcasing the effectiveness of the control strategy.

The corresponding rotor inverter waveforms, such as the rotor circuit voltage (VDFIG ROTOR) and current (IDFIG_ROTOR), are displayed in Figure 3.11(b). Figure 3.11(c) infers that the real power supplied to the rotor (PDFIG ROTOR) is inversely proportional to NSCIG. In other words, as NSCIG decreases, the DFIG rotor consumes more absolute power and vice versa. The reactive power requirement of the DFIG rotor (QDFIG-ROTOR) varies based on NSCIG, as shown in Figure 3.11(c). With an increase in NSCIG, the reactive power support for the SCIG (QSCIG) also increases. To meet this reactive power demand, the reactive power introduced at the DFIG rotor increases, leading to a subsequent increase in the current drawn from the DC link (IROTOR INV), as depicted in Figure 3.11(d). It can be substantiated that the DC currents (IBOOST, IBATT and IROTOR INV) satisfy Kirchhoff's current law, ensuring the consistency and validity of the simulation result.

Up to this point, the operation of the DFIG controller has been elucidated through its workings in the rotor circuit. The subsequent effects inside the stator side of the DFIG in the recommended hybrid system are premeditated and described. The terminal voltage values at the SCIG and the load are identical to the terminal voltage of the DFIG stator. Therefore, to demonstrate the direct power distribution between the DFIG and SCIG, merely the current variation waveforms are provided in addition to the waveform of load current in Figure 3.12. It is perceived that as NSCIG rises from 1,520 to 1,590 rpm at t = 3s, there is a rise in SCIG terminal current, ISCIG. In this interval, the current at the DFIG stator terminal, IDFIG_ST, is reduced to balance the load requirement. As mentioned earlier, this adjustment in real power provided by the DFIG is manifested in the rotor inverter MI. Conversely, when NSCIG decreases from 1,590 rpm to 1,550 rpm at t = 7s, ISCIG also decreases, increasing IDFIG_ST to meet the load requirement. The power supplied/absorbed by the DFIG and SCIG and the power absorbed by the load are shown in Figure 3.13.

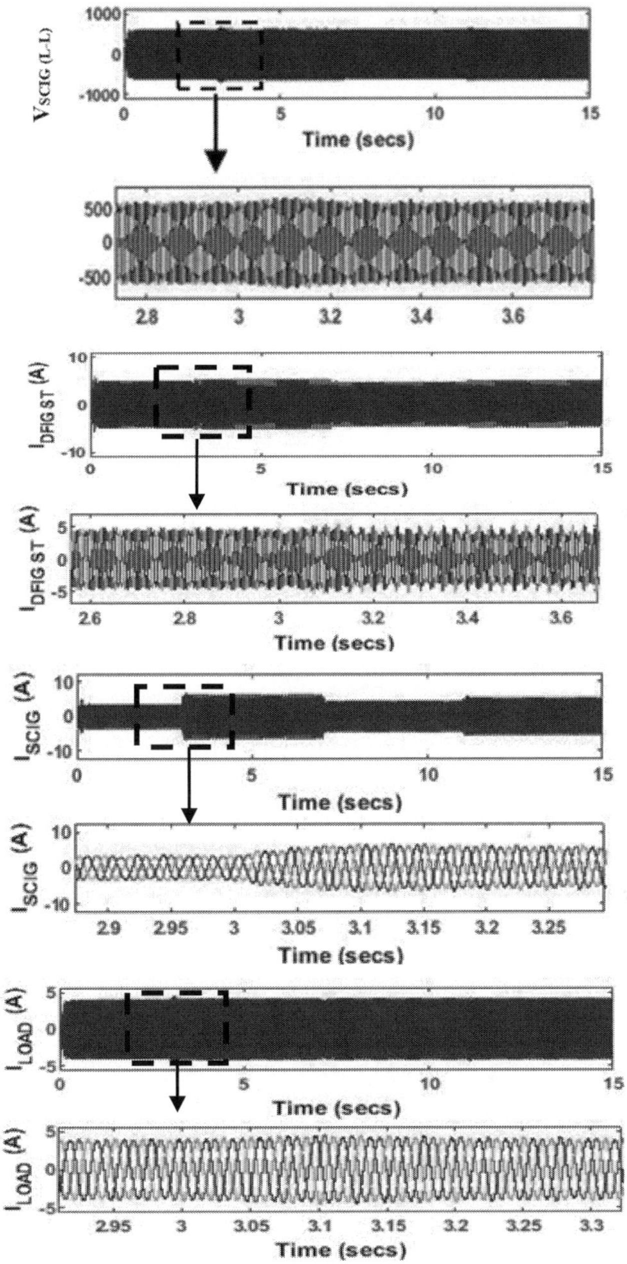

Figure 3.12. *Voltage and current waveforms of three-phase at the DFIG stator, SCIG stator and load. For a color version of this figure, see www.iste.co.uk/sundaram/energysystems.zip*

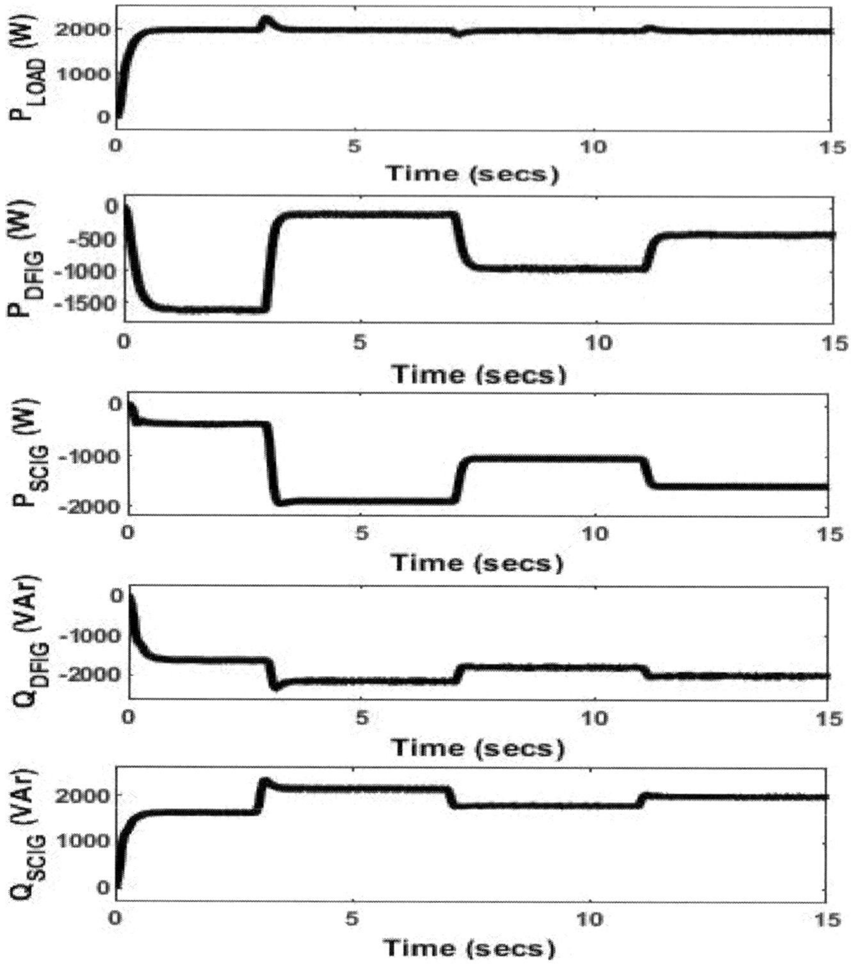

Figure 3.13. *Active and reactive power at the load, DFIG stator and SCIG stator*

Figures 3.13(a) and (b) show that with increased speed of operation, the SCIG delivers additional power to the load (Figure 3.13(b)), consequently reducing the actual power delivered by the DFIG. Additionally, the reactive power needed for SCIG to operate is powered by DFIG, that is, QSCIG = QDFIG, as in Figure 3.13(c). It is inferred that the DFIG can generate accurate and reactive power independently. It is concluded that changes in the value of irradiation do not affect the PCC voltage, and therefore, the power-sharing between the DFIG and SCIG. While the PV battery system forms the DC link of the DFIG rotor inverter, the DC side waveforms are not presented for brevity.

3.5.2. *Speed variation in DFIG*

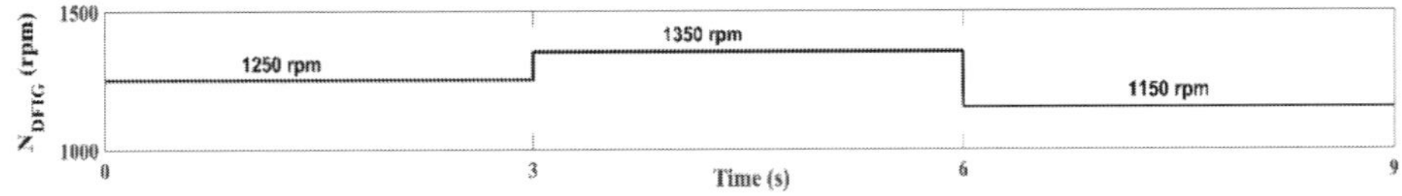

a) Set point for DFIG rotor speed

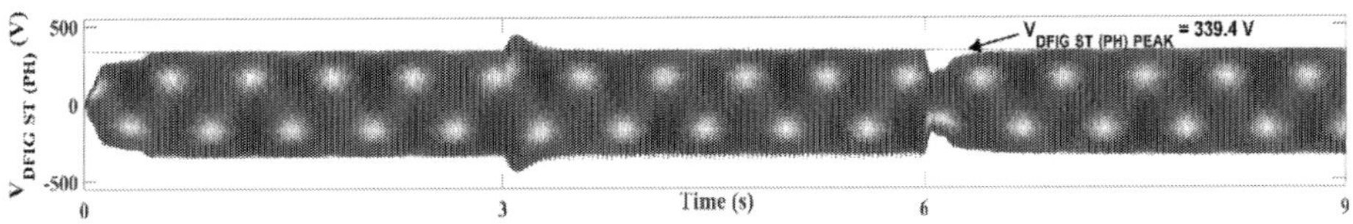

b) DFIG stator terminal voltage

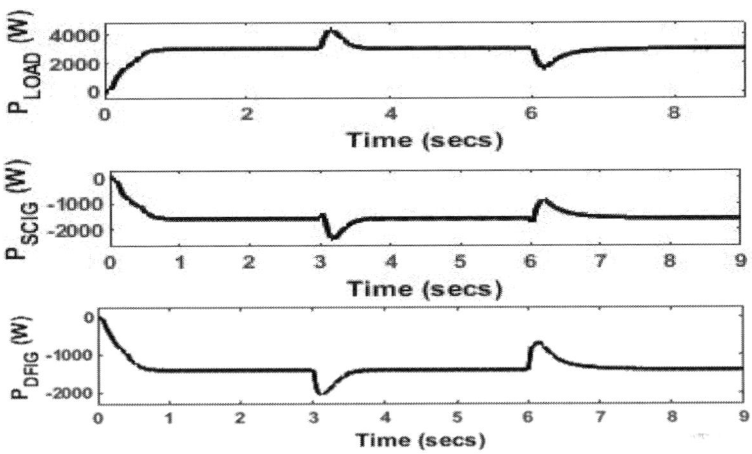

c) Active power-sharing in DFIG and SCIG

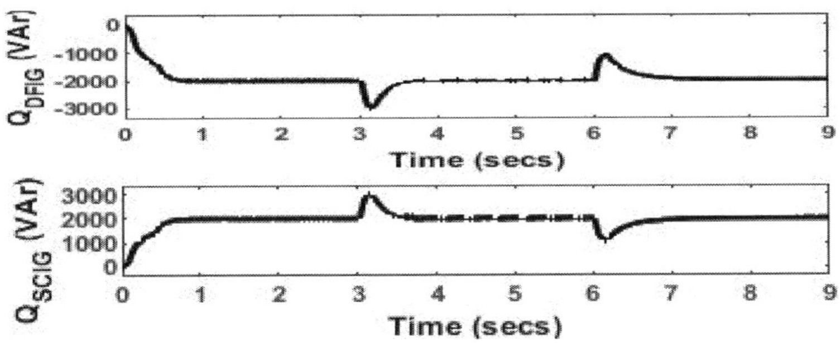

d) Reactive power balance in the AC hybrid system

Figure 3.14. *Simulation results of the hybrid system for DFIG speed variation*

In this operational mode, the DFIG is subjected to run at various speeds while the SCIG is preserved at a constant speed of 1,575 rpm. The solar PV intensity values are set as in section 3.5.2. The speed setting values for DFIG are in Figure 3.14(a), while the connected load is 3 kW. Figure 3.14(b) illustrates the stator terminal voltage of DFIG. The DFIG stator terminal voltage and frequency are regulated at 415 V and 50 Hz, respectively, achieved by applying a voltage to the rotor at the slip frequency. Figure 3.14(c) illustrates the distribution of real power shared between the SCIG and DFIG to supply the connected 3 kW load. Additionally, Figure 3.14(d) demonstrates the provision of required reactive power support from the DFIG to the SCIG.

3.5.3. *Load variation*

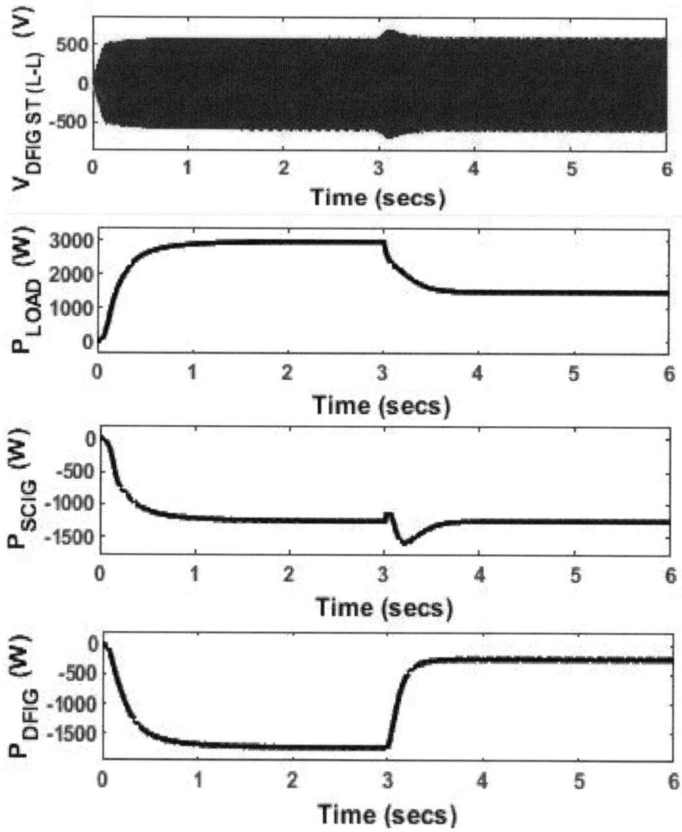

a) Voltage and real power-sharing by DFIG and SCIG

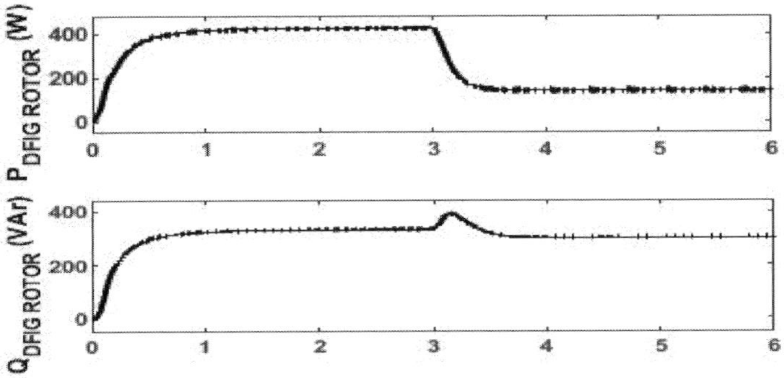

b) Real and reactive power injected into the DFIG rotor

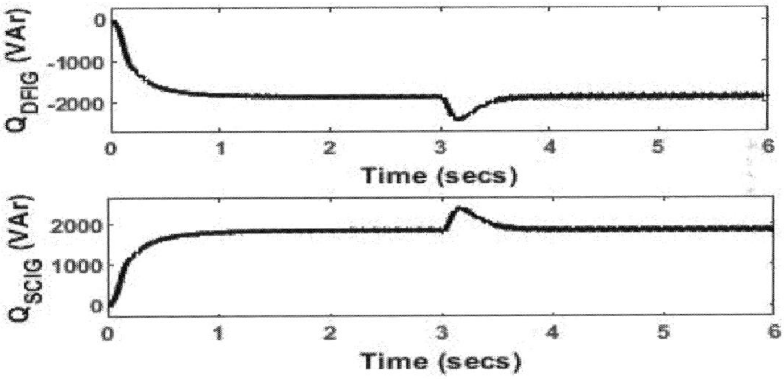

c) Reactive power of DFIG and SCIG

Figure 3.15. *Simulation results of the hybrid system for load variation*

The validation of controller performance for sudden changes in load is achieved by maintaining the speed of the DFIG, denoted as N_{DFIG}, at 1,300 rpm, while the speed of the SCIG, denoted as N_{SCIG}, is kept at 1,560 rpm. The intensity of solar PV varies continuously, while the load is adjusted from 3 to 1.5 kW at time t = 3s. Since the output power of the SCIG is not constrained, it will deliver a constant power corresponding to its operating speed. Therefore, the power generated from the DFIG will dynamically adapt to accommodate the load demand, as depicted in Figure 3.15(a). Voltage is also shown in Figure 3.15(a). As soon as the load decreases from 3 to 1.5 kW, the DFIG rotor voltage and current values will decrease accordingly. This alteration influences the real and reactive power of the DFIG rotor, resulting in a decrement, as depicted in Figure 3.15(b). It is noticeable that the alteration in actual power demand at the stator side mirrors that of the DFIG rotor power. However, since the reactive power demand at the stator side remains constant, the variation in rotor reactive power is negligible. With a constant SCIG speed, the reactive power needed by the SCIG is fulfilled by the DFIG, as illustrated in Figure 3.15(c).

3.5.4. *Wind, PV and biogas generators supply an isolated load, with real and reactive power support from the DSTATCOM for the load, and additional reactive power from the three-phase capacitor bank*

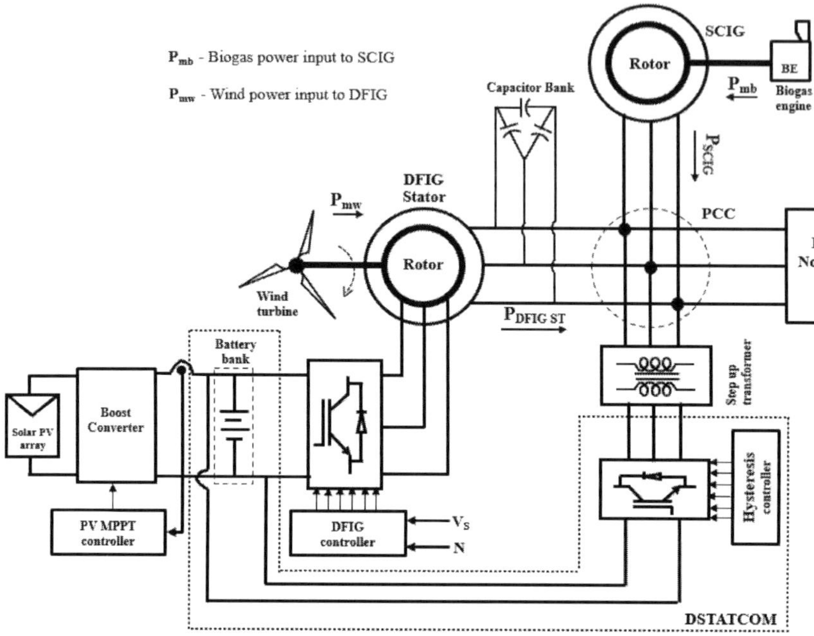

Figure 3.16. *Schematic diagram of the anticipated hybrid system with three renewable sources and DSTATCOM*

In this operating mode, the three-phase DSTATCOM module terminals are coupled at the point of PCC within the hybrid power system as depicted in Figure 3.16. The DC link for both the DFIG rotor-side inverter and the DSTATCOM is provided by the PV-battery system. The DSTATCOM is configured to deliver a fraction of the active power demand PLOAD and the reactive power prerequisite of the load QLOAD. Reactive power aid for the SCIG is supplied via a delta-connected 2 kVAr capacitor bank. Figure 3.17 illustrates active and reactive power equilibrium within the proposed scheme. The SCIG maintains a steady power output of 1.4 kW both prior to and following the closure of the DSTATCOM breaker. The DFIG provides the leftover 1.96 kW needed to fulfill the load power demand. Nonetheless, QDFIG is nearly zero as QLOAD is equivalent to zero, and QSCIG = QCAP. Once the breaker is closed at t = 2.1 s, the real power injected by DSTATCOM, $P_{STATCOM}$ is close to 50% of P_{LOAD}, that is, 1.46 kW and the reactive power introduced by DSTATCOM, $Q_{STATCOM}$ is zero, as the load is a purely resistive one. Now, PDFIG decreases to –490 W to preserve the system's power balance. In the event of lagging p.f. loads, as observed $Q_{STATCOM} = Q_{LOAD}$, $P_{STATCOM} = 0.5\ P_{LOAD}$, $P_{SCIG} = 1.04$ kW and $P_{DFIG} = P_{LOAD} - P_{SCIG} - P_{STATCOM}$. Nevertheless, QDFIG ≈ 0, after DSTATCOM is coupled in the scheme.

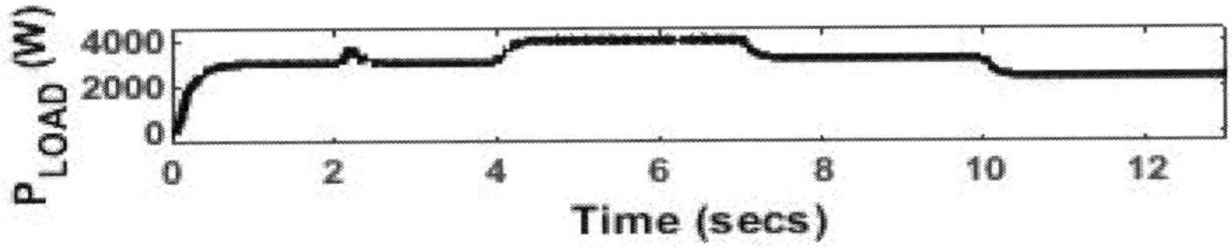

a)

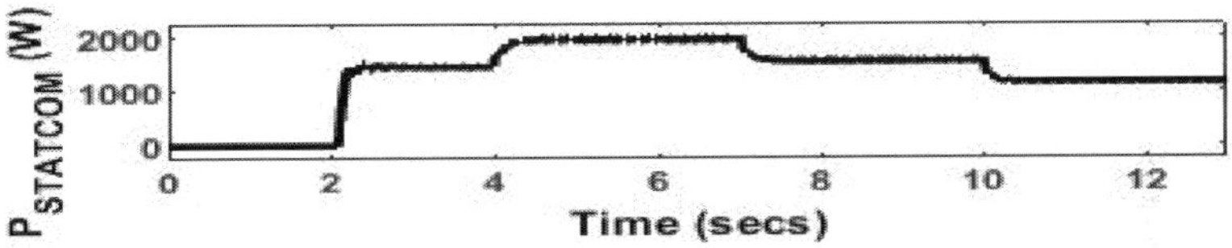

b)

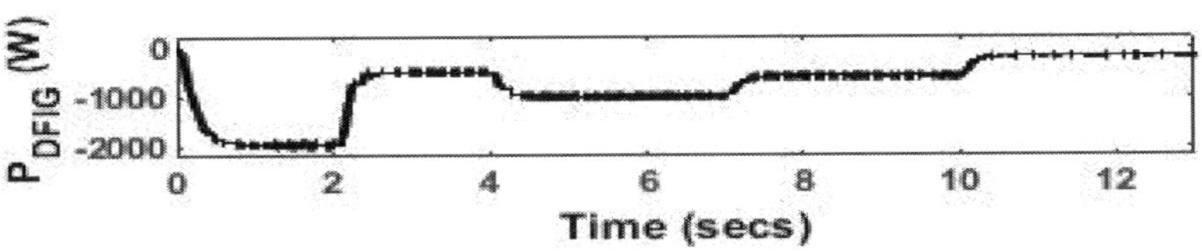

c)

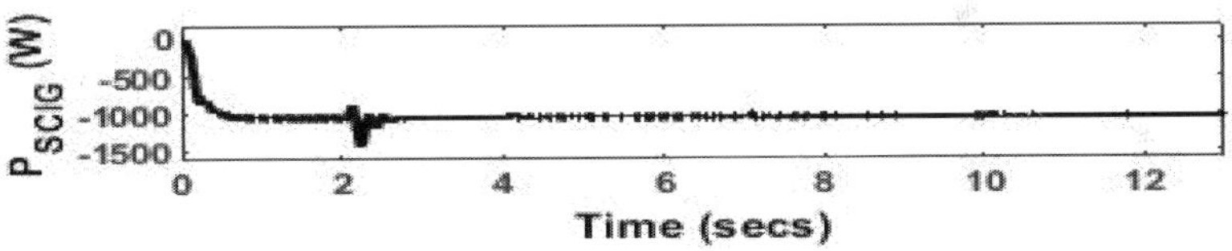

d)

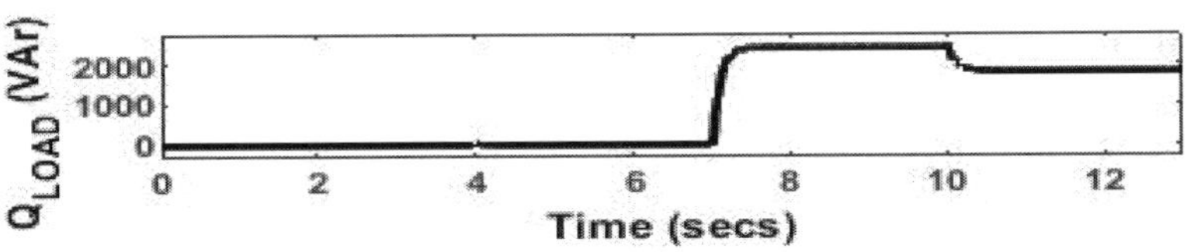

e)

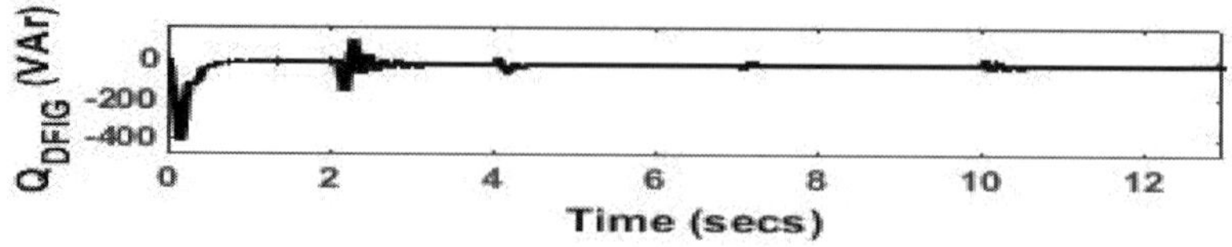

f)

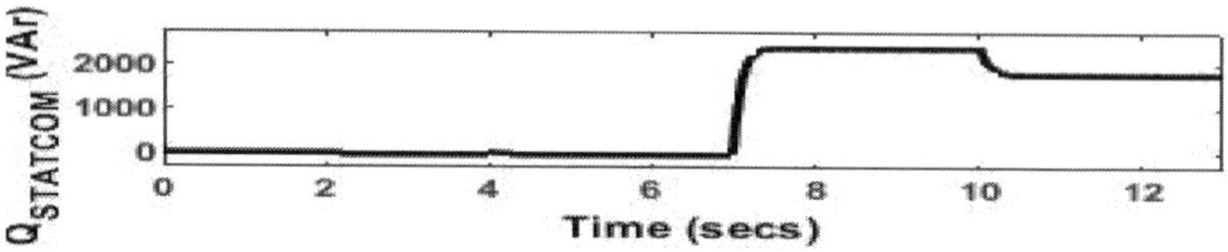

g)

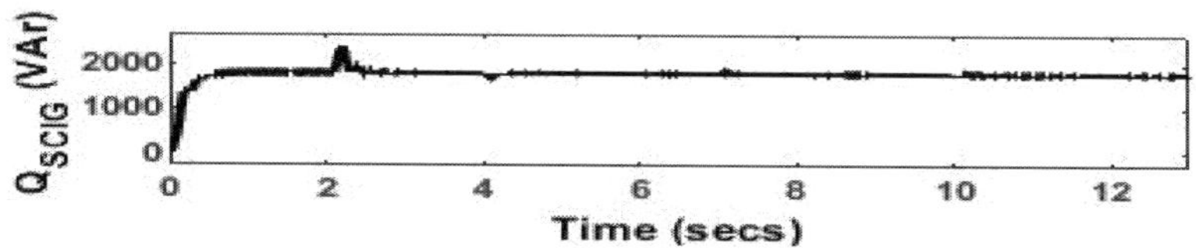

h)

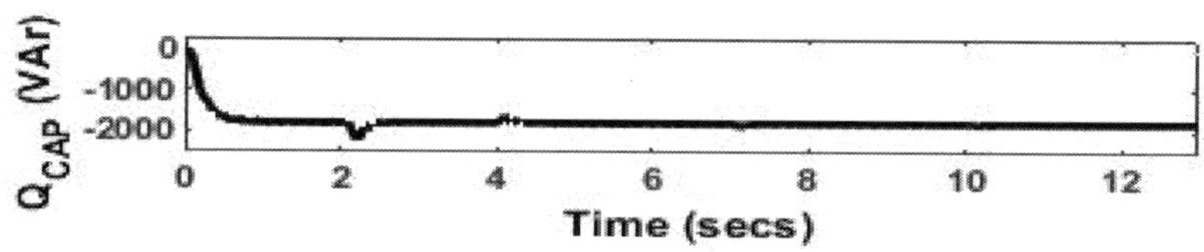

i)

Figure 3.17. *Active and reactive power distribution before and after the DSTATCOM breaker is engaged*

As depicted in Figure 3.18, the hybrid power system is powered by wind, biogas and solar energy sources. Despite wind power generation being zero or the wind speed being low, the AC loads can still be supplied by the solar PV panels if the three-phase SPWM inverter operates at 50 Hz. System simulations have been conducted, and the results demonstrate the operation of the 50 Hz inverter without the biogas generator (SCIG) as shown in Figure 3.18. Reactive power support for the SCIG is supplied via a delta-connected 2 kVAr capacitor bank and inverter. Simulation results demonstrate the scheme powering three-phase isolated loads ranging from 0.5 to 2 kW at unity power factor (UPF). The 50 Hz inverter maintains a constant AC voltage of 415 V (VL-L) for the load, controlled by a PI controller with manually tuned parameters: Kp = 0.09 and Ki = 12.

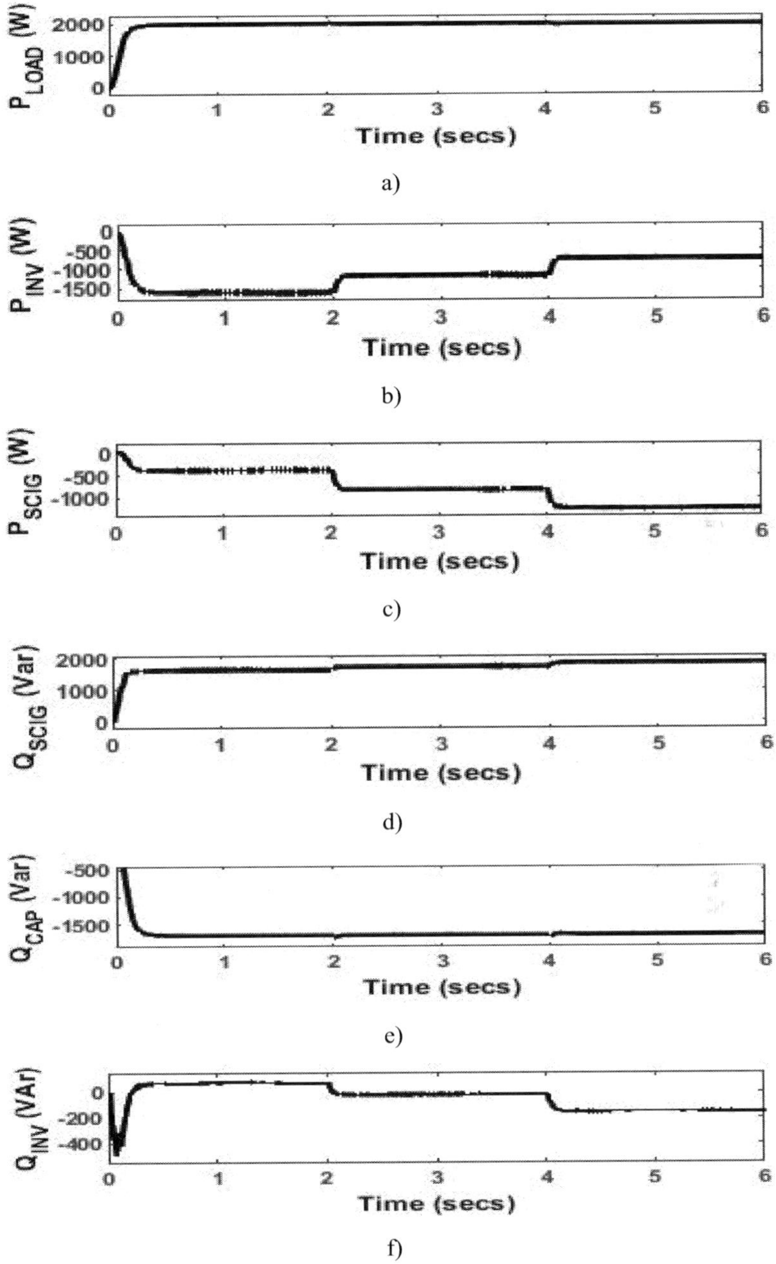

Figure 3.18. *Waveforms of load, 50 Hz inverter, capacitor and SCIG: (a) Load real power. (b) Inverter real power. (c) SCIG real power. (d) SCIG reactive power. (e) Capacitor reactive power. (f) Inverter reactive power*

3.6. Experimental outcomes of the proposed hybrid system

Tests were carried out to assess the AHGS illustrated in Figure 3.19 across different operational scenarios. Moreover, a closed-loop control tactic was employed to sustain the stator voltage of the DFIG at approximately 415 V and a frequency of roughly 50 Hz throughout all operational states. Variations in electrical parameters at various points, including the solar PV panels, boost converter, battery pack, DC link of the inverter, DFIG and biogas engine terminals, were monitored for every scenario.

3.6.1. *Experimentation utilizing solar PV panels*

The solar PV panels were optimized to function at their peak power point through the implemented COCB-MPPT control method. The solar PV system, including the DFIG, underwent laboratory testing under actual operating conditions. Details about PV panel ratings and connections are listed in Table II. Figures 3.20(a) and 3.20(b) display waveforms captured at the boost converter's input and output, showcasing gating pulses and the execution of the MPPT algorithm, including its turnaround time. Figure 3.20(c) presents steady-state test results, illustrating boost converter duty ratio variation, solar PV panel output power, the power provided to the battery under MPPT conditions, and corresponding voltage and current variations. The boost converter maintains an almost constant output voltage of 220 V while adjusting the output current to achieve MPPT conditions.

Figure 3.19. *Experimental setup of the hybrid system. For a color version of this figure, see www.iste.co.uk/sundaram/energysystems.zip*

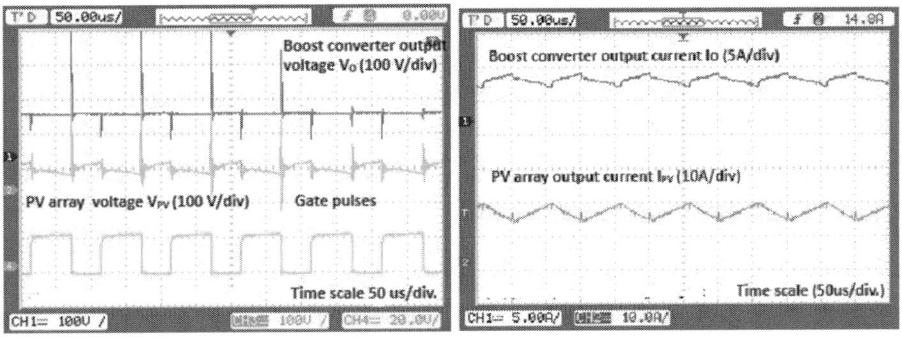

a) Steady-state voltage and current waveforms

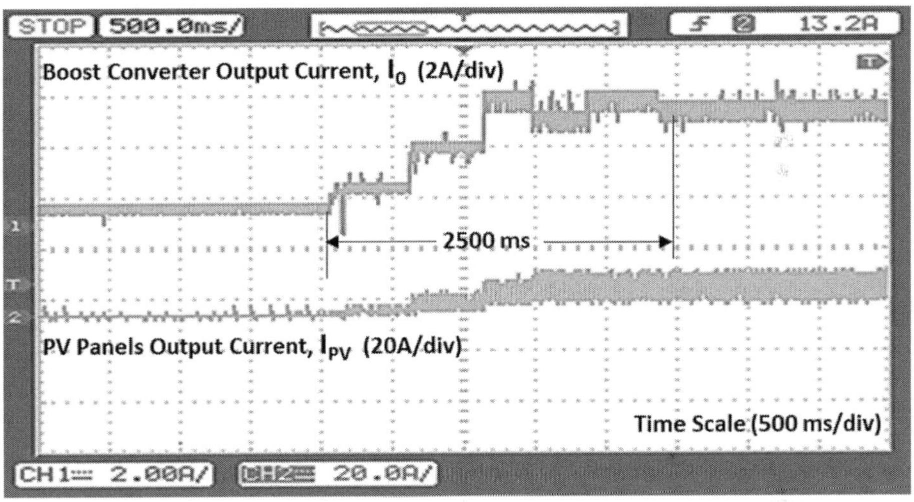

b) The dynamic response of the MPPT algorithm

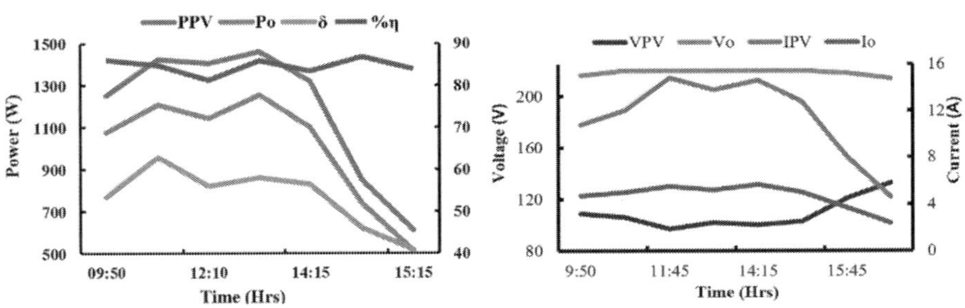

c) Test outcomes derived from a solar PV system

Figure 3.20. *Test outcomes derived from the solar PV system. For a color version of this figure, see www.iste.co.uk/sundaram/energysystems.zip*

Solar PV array voltage fluctuates between 95 and 133 V, accompanied by equivalent current variations. Figure 3.20 demonstrates the MPPT algorithm's ability to optimize power extraction from solar PV panels by adjusting the converter duty ratio in a closed-loop manner.

3.6.2. *Experimentation utilizing solar PV panels and DFIG system*

From the experimental data presented in Figure 3.21, it is marked that the power produced by the solar PV panels is utilized both in charging the battery and providing power to the rotor of DFIG during periods of high irradiation. This is illustrated in Figure 3.21(a) with the load at stator terminals below 2 kW and from Figure 3.21(b) for all loads, Io = IB + Iinv.

During periods of low irradiation, the power generated by the solar PV panels may not meet the demand at the terminals of the DFIG rotor. Consequently, the battery delivers additional power beyond the solar PV output, as depicted in Figure 3.22(a) for loads exceeding 2 kW and illustrated in Figure 3.21(c) for all the loads, (Io + IB) = Iinv.

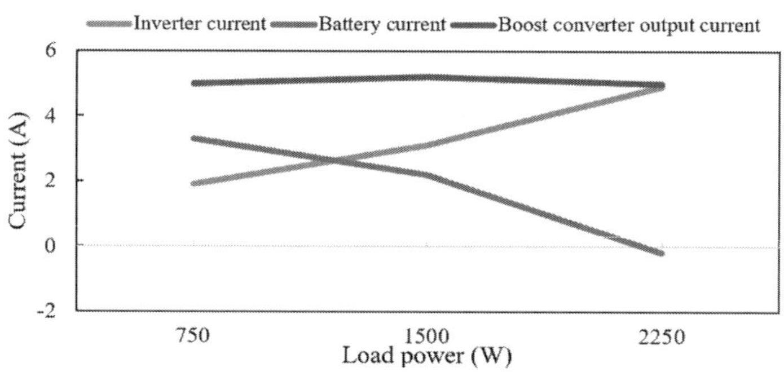

a) Test conducted at 11:00 and DFIG speed = 1,150 rpm

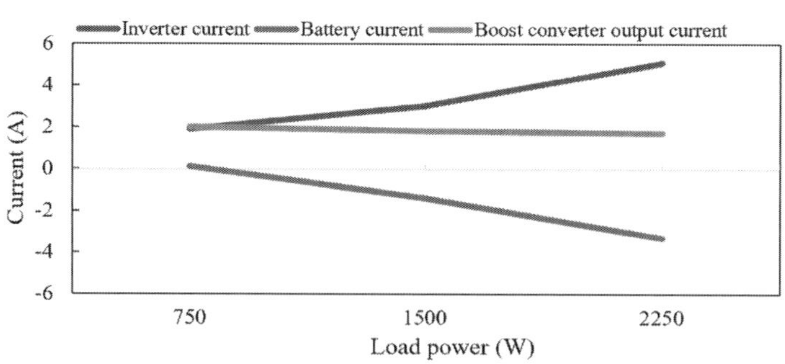

b) Test conducted at 13:00 and DFIG speed = 1,350 rpm

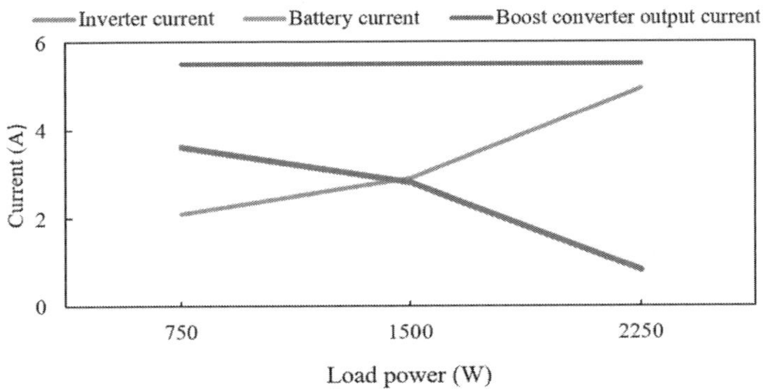

c) Test conducted at 17:00 and DFIG speed 1,250 rpm

Figure 3.21. *Wind-solar hybrid system performance assessment. For a color version of this figure, see www.iste.co.uk/sundaram/energysystems.zip*

When solar PV output is zero, the battery provides power to the DFIG rotor for a period determined by the battery capacity. In this scenario, the battery current equals the DC link current of the inverter (IB = Iinv). Therefore, battery charging or discharging relies on excess or deficiency power from solar PV panels for wind-solar hybrid systems.

3.6.3. *Integrated operation of solar, wind and biogas sources*

The AHGS undergoes testing in the laboratory alongside the wind-driven DFIG and the biogas engine-driven SCIG, covering a range of load configurations. Two induction machines with identical specifications have been utilized: a three-phase, four-pole, 415 V, 50 Hz, three hp, star-connected stator. As previously mentioned, a DC motor was a prime mover for running the DFIG. Correspondingly, the SCIG was connected to a separately excited DC machine to emulate a biogas engine. Experimental trials were conducted by applying different loads at the PCC while simultaneously running the DFIG and the SCIG at varying speeds. This comprehensive approach allowed for a thorough examination and assessment of the performance characteristics of both generator systems under diverse operational conditions. The variations in the terminal voltage, current and power of the stator and rotor of the DFIG in different load and speed conditions are depicted in Figures 3.22(a) and (b). The test results shown in Figure 3.22 indicate that the DFIG control scheme consistently preserves the stator terminal voltage at 415 V and a frequency of 50 Hz under all working conditions.

Moreover, the power delivered to the rotor rises with an increase in load at a constant speed. Conversely, for a constant load, the rotor power of DFIG reduces with the rise in speed. During this experiment, reactive power demanded by DFIG was exclusively fulfilled by the battery-SPWM inverter system on the rotor side.

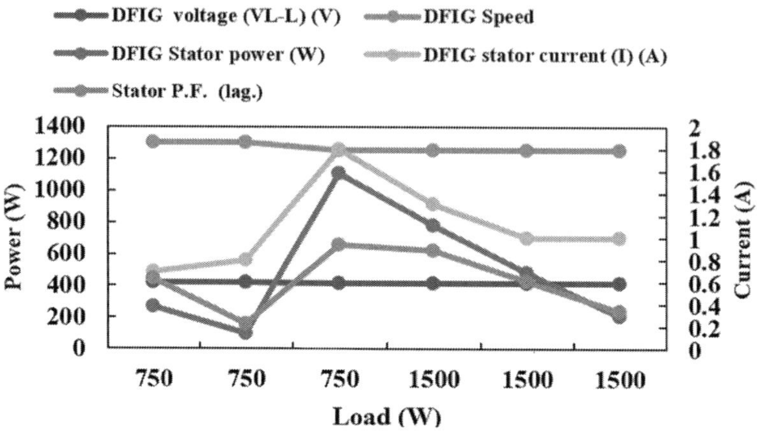

a)

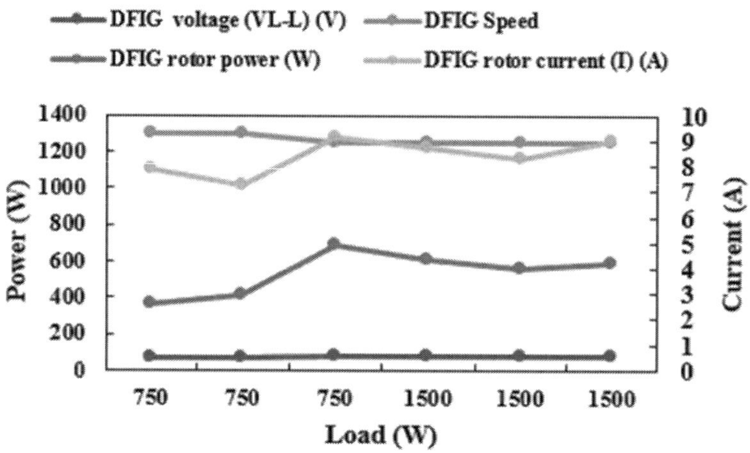

b)

Figure 3.22. *Variation of DFIG stator and rotor parameters. For a color version of this figure, see www.iste.co.uk/sundaram/energysystems.zip*

The distribution of the total load at the PCC between the DFIG and the SCIG is evident from Figure 3.23(g). Furthermore, the power supplied by the SCIG increases proportionally with its rotational speed. To maintain power balance, the power supplied by the DFIG decreases, even under constant speed operation. Voltage and current waveforms were captured and are presented for three distinct operating conditions in Figures 3.23(a)–(f). The recorded electrical quantities are detailed and indicated.

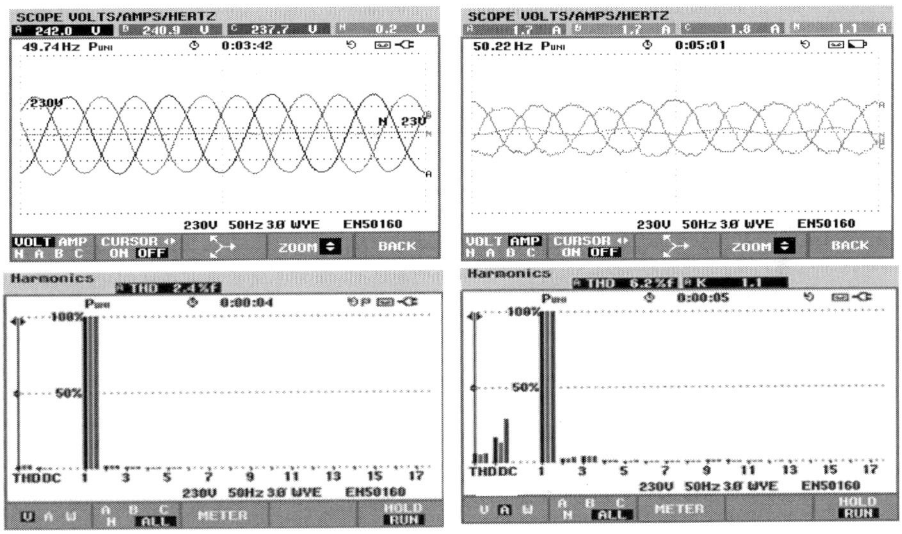

a) DFIG stator phase voltage and current with THD

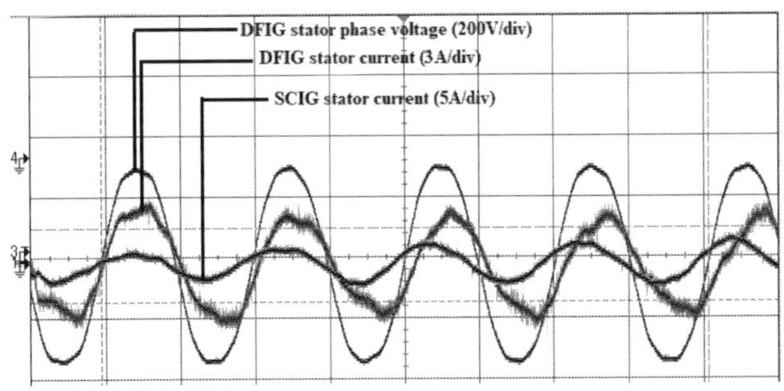

b) DFIG and SCIG waveforms for a load of 1.5 kW and at 1,250 rpm (SCIG speed 1,515 rpm)

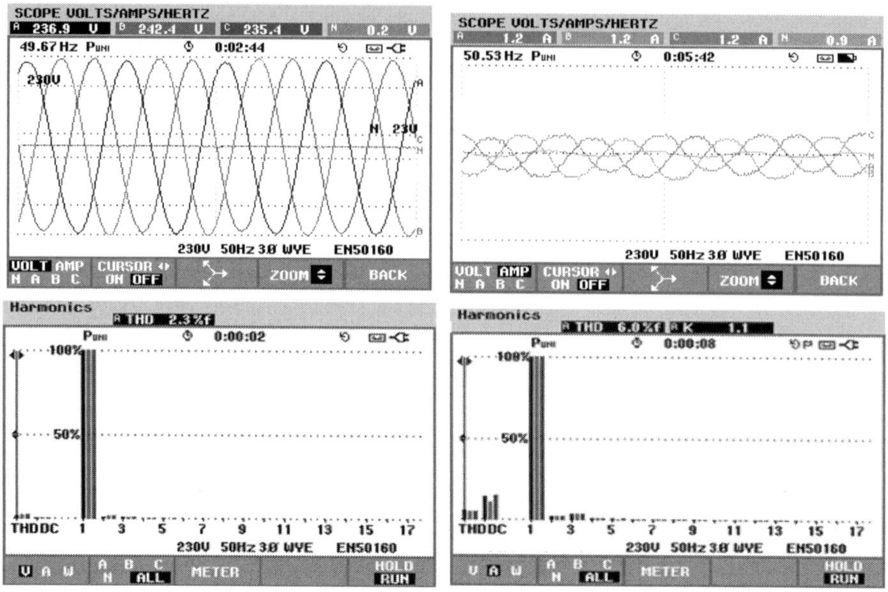

c) DFIG stator phase voltage and current with THD, respectively

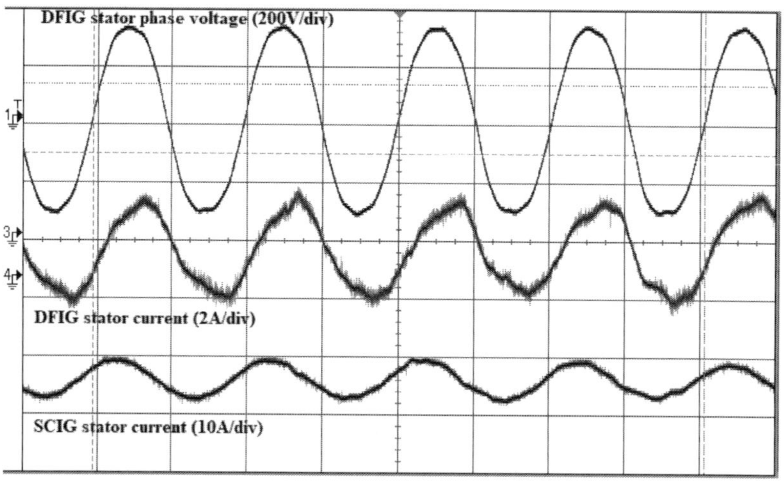

d) DFIG and SCIG waveforms for a load of 1.5 kW and at 1,250 rpm (SCIG speed 1,530 rpm)

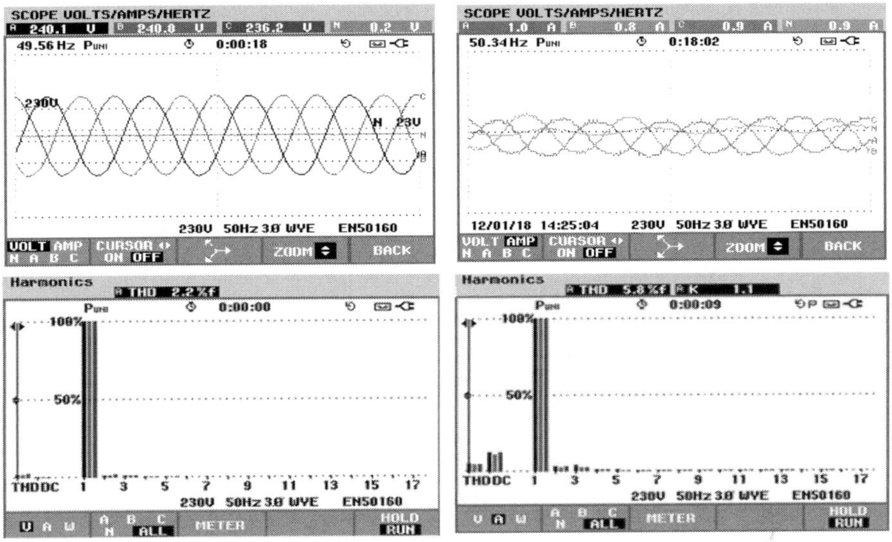

e) DFIG stator phase voltage and current with THD, respectively

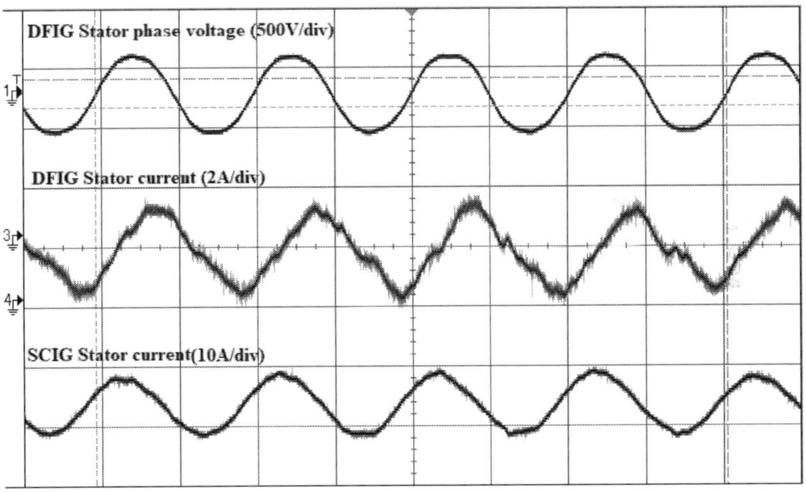

f) DFIG and SCIG waveforms for a load of 1.5 kW and at 1,250 rpm (SCIG speed 1,570 rpm)

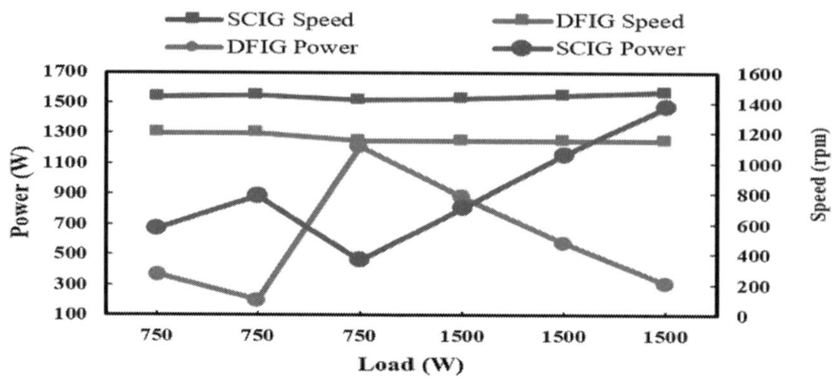

g) Load sharing by DFIG and SCIG

Figure 3.23. *Experiment results from the proposed AHRGS for different load and speed settings. For a color version of this figure, see www.iste.co.uk/sundaram/energysystems.zip*

Table 3.3 compares the THD voltage results of this study with those from existing research. Notably, previous studies did not integrate PV battery, wind and biogas energy with battery storage in a microgrid system. The VFCL controller in this study achieved the lowest voltage THD value compared to others. Specifically, the voltage THD value differed by only 0.45% from the PI control method with PV-wind-biogas integration. Additionally, when compared to the method in Badruhisham et al (2022), this study demonstrated a higher THD voltage value.

Method	Energy sources	Voltage THD (%)
Proposed Method	PV-Wind-Biogas	2.30
PI (Badruhisham et al. 2022)	PV-Wind-Biogas	1.85
Voltage source converter (Shan et al 2019)	PV-Wind	3.80
Proportional resonance and predictive current (Alhamrouni et al. 2022)	PV-Wind	3.15
PI (Guo et al. 2022)	PV-Wind	3.20

Table 3.3. *Evaluating outcomes from existing articles*

3.7. Conclusion

This study presents a pioneering AHRGS that effectively integrates wind, solar, biogas and battery storage technologies to deliver reliable energy to isolated rural communities. The innovative system design incorporates a solar PV array connected to the rotor side of the DFIG via a converter-battery-inverter configuration. The developed VFCL control strategy was executed using an FPGA (Altium Nanoboard 3000) controller. Experimental substantiation was conducted on a three-phase, four-pole, 415 V, 50 Hz, 3-HP, star-connected DFIG, with the reference stator voltage set at 415 V and 50 Hz. The controller established adequate performance in tracking the set reference stator voltage and frequency under various operating conditions.

The development of a closed-loop control system for the solar PV boost converter, featuring an efficient MPPT algorithm based on the COCB method, highlights the system's operational efficiency. This algorithm only requires monitoring the output current of the boost converter, simplifying the control process while maximizing energy extraction from the solar PV array.

Extensive experimentation has validated the seamless integration of the wind-solar hybrid system, which adeptly manages the fluctuations in solar power output and DFIG rotor power by strategically storing or drawing energy from the battery. This capability ensures a stable and reliable power supply for remote communities, enabling them to harness their abundant wind and solar resources effectively.

The proposed hybrid system has been tested under diverse operational scenarios, demonstrating its ability to maintain optimal performance across varying wind and solar conditions. By employing a VFCL controller alongside a wind-driven DFIG, the system adapts to the availability of renewable resources and battery capacity, guaranteeing a continuous energy supply to isolated loads. Test results indicate that the VFCL control strategy successfully maintains the stator voltage at 415 V and the frequency at 50 Hz, each surrounded by a ±1% tolerance, in all working conditions. Experimental outcomes indicate that the measured THD values remain below 3% for voltage and below 6.5% for current. This demonstrates that the hybrid solar PV–wind–biogas microgrid maintains stability, particularly in distribution lines.

The comprehensive analysis of voltage, current and both active and reactive power waveforms further shows the efficacy of the AHRGS. These findings emphasize the critical importance of well-designed control strategies in optimizing hybrid renewable energy systems. The successful implementation of this system not only addresses the energy needs of remote communities but also contributes to the broader goal of increasing renewable energy adoption in India, positioning this research as a significant step toward sustainable energy development in underserved regions.

3.8. Acknowledgments

The authors gratefully acknowledge the support provided by the authorities of B.M.S. College of Engineering, Bengaluru, India, through the FRPS grant (R&D/FRPS/2023-24/EE/01).

3.9. References

Al-Ammar, E.A., Habib, H.U.R., Kotb, K.M., Wang, S., Ko, W., Elmorshedy, M.F., Waqar, A. (2022). Residential community load management based on optimal design of standalone HRES with model predictive control. *IEEE Access*, 8, 12542–12572.

Alhamrouni, I., Wahab, W., Salem, M., Rahman, N.H.A., Awalin, L. (2019). Modeling of micro-grid with the consideration of total harmonic distortion analysis. *Indones. J. Electr. Eng. Comput. Sci.*, 15(2), 581–592.

Ali, M.F., Sheikh, M.R.I., Mamun, A.A., Hossen, M.J. (2025). Techno-economic, predictive modeling, and demand response analysis of a renewable energy-based microgrid for residential applications. *IEEE Access*, 13, 53748–53771.

Badruhisham, S.H., Abu Hanifah, M.S., Yusoff, S.H., Hasbullah, N.F., Yaacob, M. (2022). PI controller for hybrid biomass-solar photovoltaic-wind in microgrid: A case study of Mersing, Malaysia. *IEEE Access*, 10, 95151–95160.

Chauhan, A. and Saini, R.P. (2013). Renewable energy-based power generation for stand-alone applications: A review. In *Proc. Int. Conf. Energy Efficient Technol. Sustain.*, 424–428.

Diab, A.A.Z., Sultan, H.M., Mohamed, I.S., Kuznetsov, O.N., Do, T.D. (2019). Application of different optimization algorithms for optimal PV/wind/diesel/battery storage stand-alone hybrid microgrid sizing. *IEEE Access*, 7, 119223–119245.

Fernández-Muñoz, D. and Pérez-Díaz, J.I. (2021). Unit commitment in a hybrid diesel/wind/pumped-storage isolated power system considering the net demand intra-hourly variability. *IET Renew. Power Gener.*, 15(1), 30–42.

Fernández-Muñoz, D. and Pérez-Díaz, J.I. (2023). Optimisation models for the day-ahead energy and reserve self-scheduling of a hybrid wind–battery virtual power plant. *J. Energy Storage*, 57, 106296.

Guo, X.-H., Chang, C.-W., Chang-Chien, L.-R. (2022). Digital implementation of harmonic and unbalanced load compensation for voltage source inverter to operate in grid forming microgrid. *Electronics*, 11(6), 886.

Hassan, Q., Algburi, S., Sameen, A.Z., Salman, H.M., Jaszczur, M. (2023). A review of hybrid renewable energy systems: Solar and wind-powered solutions: Challenges, opportunities, and policy implications. *Results Eng.*, 20, 101621.

Hassan, Q., Viktor, P., Al-Musawi, T.J., Ali, B.M., Algburi, S., Alzoubi, H.M., Al-Jiboory, A.K., Sameen, A.Z., Salman, H.M., Jaszczur, M. (2024). The renewable energy role in the global energy transformations. *Renewable Energy Focus*, 48, 100545.

Hoseinzadeh, S. and Astiaso Garcia, D. (2022). Techno-economic assessment of hybrid energy flexibility systems for islands' decarbonisation: A case study in Italy. *Sustain. Energy Technol. Assess.*, 51, 101929.

Javaid, S., Kaneko, M., Tan, Y. (2024). Energy balancing of power system considering periodic behavioral pattern of renewable energy sources and demands. *IEEE Access*, 12, 70245–70262.

Kant, K., Jain, C., Singh, B. (2017). A hybrid diesel-wind-PV-based energy generation system with brushless generators. *IEEE Trans. Ind. Inform.*, 13(4), 1714–1722.

Manyonge, A.W., Ochieng, R.M., Onyango, F.N., Shichikha, J.M. (2012). Mathematical modelling of wind turbine in a wind energy conversion system: Power coefficient analysis. *Applied Mathematical Sciences*, 6, 4527–4536.

Moosavian, S.F., Noorollahi, Y., Shoaei, M. (2024). Renewable energy resources utilization planning for sustainable energy system development on a stand-alone island. *J. Clean. Prod.*, 439, 140892.

Pérez, A.E., Zamora Gomez, E., Polupan, G. (2025). Analysis and study of complete and incomplete combustion in an internal combustion engine using different amounts of biogas. *IEEE Latin Am. Trans.*, 23(5), 397–404.

Puchalapalli, S. and Singh, B. (2020). A single input variable FLC for DFIG-based WPGS in standalone mode. *IEEE Trans. Sustain. Energy*, 11(2), 595–607.

Sam, K., Kumaresan, N., Gounden, N. (2017). Wind-driven stand-alone DFIG with battery and pumped hydro storage system. *Sādhanā Acad. Proc. Eng. Sci.*, 42(2), 173–185.

Sekhar, N. and Kumaresan, N. (2022). Operation and control a stand-alone power system with integrated multiple renewable energy sources. *Wind Eng.*, 46(1), 221–239.

Sekhar, N., Kumaresan, N., Gounden, N. (2019). Experimental investigation on a new hybrid system employing wind-driven DFIG and solar PV panels. *J. Inst. Eng. India Ser. B.*, 100, 1–14.

Sekhar, N., Tatabhatla, V.M.R., Kumaresan, N. (2025). A novel single sensor maximum power point tracking algorithm for standalone photovoltaic systems. *Int. J. Model. Simul.*, 1–20. doi: 10.1080/02286203.2025.2485441.

Shan, Y., Hu, J., Chan, K.W., Fu, Q., Guerrero, J.M. (2019). Model predictive control of bidirectional DC-DC converters and AC/DC interlinking converters: A new control method for PV-wind-battery microgrids. *IEEE Transactions on Sustainable Energy*, 10(4), 1823–1833.

Sharma, R.K., Mudaliyar, S., Mishra, S. (2018). Power management and economic load dispatch-based control of hybrid PV-battery-diesel standalone AC system. In *IEEMA Engineer Infinite Conference*. IEEE, New Delhi.

Shayan, M.E., Najafi, G., Ghobadian, B., Gorjian, S., Mazlan, M. (2023). A novel approach of synchronisation of the sustainable grid with an intelligent local hybrid renewable energy control. *Int. J. Energy Environ. Eng.*, 14, 35–46.

Silva, A.R. and Estanqueiro, A. (2020). Optimal planning of isolated power systems with near 100% of renewable energy. *IEEE Trans. Power Syst.*, 35(2), 1274–1283.

Temirbekova, M., Aliyarova, M., Iliev, I., Yelemanova, A., Sagintayeva, S. (2020). The generation of a mathematical model of the biogas production process from organic municipal solid waste. *E3S Web Conf.*, 180.

Tiwari, S.K., Singh, B., Goel, P.K. (2019). Control of wind–diesel hybrid system with BESS for optimal operation. *IEEE Trans. Ind. Appl.*, 55(2), 1863–1872.

Ullah, Q.Z., Elkadeem, M.R., Kotb, K.M., Taha, I.B.M., Wang, S. (2021). Multi-criteria decision-making model for optimal planning of on/off grid hybrid solar, wind, hydro, biomass clean electricity supply. *Renew. Energy*, 179, 885–910.

Vijayakumar, K., Kumaresan, N., Ammasai Gounden, N. (2012). Operation and closed-loop control of wind-driven stand-alone DFIGs using single inverter-battery system. *IET Electr. Power Appl.*, 6(3), 162–171.

Vijayakumar, K., Kumaresan, N., Ammasai Gounden, N.G., Tennakoon, S.B. (2013a). Real and reactive power control of hybrid excited wind-driven grid-connected DFIG. *IET Power Electron.*, 6(6), 1197–1208.

Vijayakumar, K., Kumaresan, N., Ammasai Gounden, N. (2013b). Operation of inverter assisted wind-driven slip ring induction generator for stand-alone power supplies. *IET Electr. Power Appl.*, 7(4), 256–269.

Wang, Y., Cai, C., Liu, C., Han, X., Zhou, M. (2022). Planning research on rural integrated energy system based on coupled utilization of biomass-solar energy resources. *Sustain. Energy Technol. Assess.*, 53(Part A), 102416.

Xu, A., Awalin, L.J., Al-Khaykan, A., Forootan Fard, H., Alhamrouni, I., Salem, M. (2023). Techno-economic and environmental study of optimum hybrid renewable systems, including PV/wind/gen/battery, with various components to find the best renewable combination for Ponorogo regency, East Java. *Sustainability*, 15(3), 1802.

Yu, G., Liu, C., Tang, B., Chen, R., Lu, L., Cui, C., Hu, Y., Shen, L., Muyeen, S.M. (2022). Short term wind power prediction for regional wind farms based on spatial–temporal characteristic distribution. *Renew. Energy*, 199, 599–612.

Yuan, J. and Weng, Y. (2024). Enhance unobservable solar generation estimation via constructive generative adversarial networks. *IEEE Trans. Power Syst.*, 39(1), 2251–2263.

Zhao, L., Li, Q., Ding, G. (2024). An intelligent web-based energy management system for distributed energy resources integration and optimization. *J. Web Eng.*, 23(1), 165–195.

4

Techno-Economic Analysis of a Hybrid Renewable Energy System with Solar PV, Wind and Biomass for Off-Grid Communities: A HOMER-Based Case Study in Aalborg, Denmark

The increasing global energy demand and depletion of conventional energy sources necessitate the development of sustainable and reliable renewable energy systems. This chapter presents a comprehensive design, optimization and feasibility assessment of an off-grid solar PV/wind/biomass/battery hybrid mini-grid for community electrification in Aalborg, Denmark, using HOMER Pro software. The proposed system integrates 60 kW solar PV, 20 kW wind turbines, 40 kW biomass gasifier and 74 kWh battery storage to ensure a reliable and cost-effective power supply. Key environmental factors, including solar irradiation, wind speed and biomass availability, are considered in the analysis.

Techno-economic performance is evaluated through levelized cost of energy (LCOE), net present cost (NPC) and loss of power supply probability (LPSP). The optimal configuration achieves LCOE of \$0.195/kWh and NPC of \$699,494, demonstrating cost-competitiveness, 100% renewable penetration, eliminating dependency on fossil fuels, minimal unmet load (0.023%) and capacity shortage (0.095%), ensuring high reliability and annual CO_2 emissions of only 42.2 kg, highlighting environmental benefits. Challenges such as excess power generation (8.56%) and battery storage losses are addressed through system sizing and control strategies.

This study provides a practical framework for deploying hybrid mini-grids in remote communities, emphasizing reliability, affordability and sustainability.

Chapter written by Venkatesh BODDAPATI, Nethravathi S, Prakash D.B., Nindra SEKHAR and Arul Daniel S.

4.1. Introduction

The need for energy is growing daily, and traditional energy sources are running out. Developing the infrastructure required to tap untapped renewable energy sources is essential. One of the best prospects for extracting renewable energy sources is developing mini-grids for different load applications, such as residential, community and industrial. Before creating a mini-grid for a particular load and location, it is essential to investigate technical and economic feasibility. Furthermore, designing an optimal microgrid architecture involves complex trade-offs between cost, reliability, scalability and environmental impact, posing significant challenges in system optimization.

Intermittency and reliability of renewable energy sources are one of the key challenges in microgrid optimization. The variable nature of solar and wind energy necessitates advanced forecasting and storage solutions to maintain a stable power supply. Also, while renewable energy costs have declined, the LCOE of hybrid systems must remain competitive with diesel-based grids. Over-sizing components increases NPC, whereas under-sizing risks load shedding (Shivanaganna et al. 2024). Further, integrating multiple energy sources, viz., PV, wind and biomass, with storage requires sophisticated energy management systems (EMS) to prioritize dispatchable sources over intermittent ones (Nethravathi and Murali 2021).

Recent research has leveraged computational tools such as HOMER, PSO and AI-based algorithms to optimize HRES (Boddapati, and Daniel 2020; https://www.homerenergy.com/products/pro/index.html).

This chapter describes a hybrid energy-based mini-grid design and feasibility assessment for a community load application (Jayanth et al. 2018; Munjer et al. 2018; Boddapati et al. 2021a, 2021b, 2021c; Boddapati and Daniel 2021; Boddapati et al. 2022a, 2022b; Manas et al. 2023). The contributions of the chapter are as follows:

– Proposing a novel hybrid architecture combining solar PV (60 kW), wind (20 kW), biomass gasifier (40 kW) and battery storage (74 kWh) – a configuration rarely explored for Nordic climates (Aalborg, Denmark).

– Quantifying reliability metrics: unlike generic feasibility studies, we evaluate LPSP (0.023%) and capacity shortage (0.095%) to ensure practical viability.

– Addressing real-world challenges: by analyzing excess power (8.56%) and biomass fuel constraints (673 tons/year), we provide actionable insights for community-scale deployment.

– Benchmarking against eight configurations: comparative analysis reveals the proposed system's superiority in LCOE ($0.195/kWh), NPC ($699k) and 100% renewable penetration.

4.2. Proposed system architecture

Solar PV, biomass, wind turbines, DG, battery storage and community load are all included in the proposed system architecture. The community load considered for the analysis is shown in Figure 4.1. The DG increases the off-grid system's dependability. An AC outlet serves a community load. AC and DC buses are used in the architecture to supply electricity, although the AC load is considered for analytical purposes. A bidirectional converter keeps the power flowing between the solar PV, wind turbine, battery and DG. The suggested system's feasibility is evaluated for the Aalborg location in Denmark.

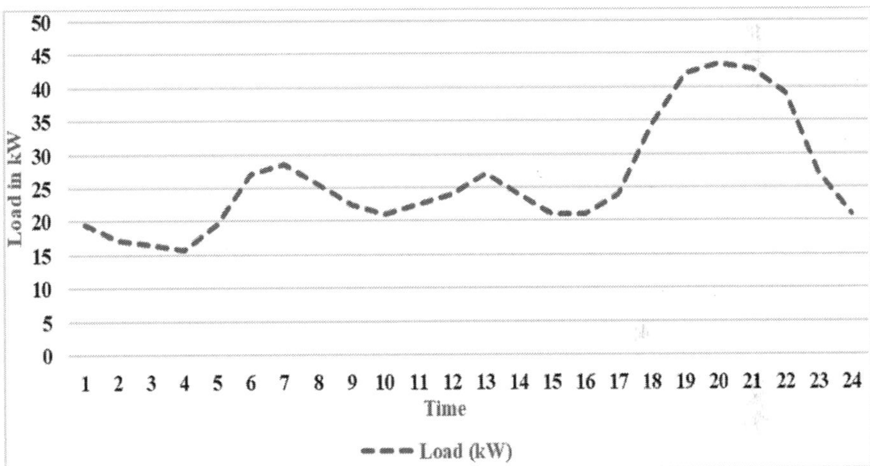

Figure 4.1. *Community load variation*

4.3. System modeling

Mathematical modeling aids in understanding the characteristics and effects of changes in each of the system's variables. The following sections describe the system modeling of the different parts of the suggested approach. The proposed mini-grid is run in off-grid mode, as seen in Figure 4.2.

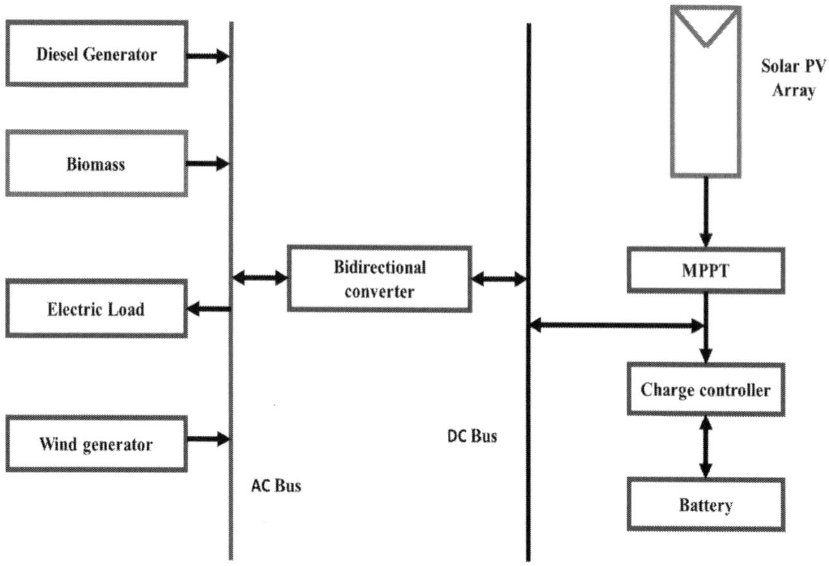

Figure 4.2. *AC mini-grid. For a color version of this figure, see www.iste.co.uk/sundaram/energysystems.zip*

4.3.1. *PV array output power*

One component of the solar PV module is a photovoltaic cell, which transforms light into electrical power. Equations and module specifications provided by the manufacturer can be used to evaluate the output power. The temperature and irradiance inputs determine the module's output. The output power and the input irradiance are considered to be proportionate. Using equation [4.1], the PV module's output power is determined (Soon and Low 2015).

$$CapCapP_{sol} = P_{rat}f_{loss}\frac{G_h}{G_s}[1 + \alpha_p(T_C - T_s)] \quad [4.1]$$

P_{rat} represents the solar PV panel's power generation capacity, while its loss is defined by f_{loss}. The losses caused by drift, temperature and partial shading are included in the loss factor. Hourly solar incidence radiation on solar panels is indicated as G_h (W/m^2). T_c is the current cell temperature in the current phase, T_s is the PV cell temperature under STC, α_p is the temperature coefficient of the power and G_s stands for standard incident irradiation (1,000 W/m^2).

4.3.2. *Wind turbine output power*

The output power of the wind turbines depends on wind speed. The electrical power output of the wind turbine is assessed using equation [4.2] (Boddapati and Daniel 2021).

$$P_e = \frac{1}{2} \times \rho \times C_p \times A \times V^3 \times 10^{-3} \qquad [4.2]$$

C_p is the wind turbine system's power coefficient, representing the air density. A indicates the surface swept by the rotor (m^2), and V denotes the wind speed (m/s).

4.3.3. *Battery modeling*

The state of charge (SOC) often determines the battery's charging time and rate. The battery's SOC or DOD can be used to characterize its charging state. The battery's SOC is 100% when fully charged. The DOD of the battery is 100% if it is completely discharged. The battery's lifespan must be extended by avoiding a 100% DOD scenario. SOC is assumed to be 20% and DOD to be 80% when calculating the potential community load's power needs. Equation [4.3] can be used to calculate the battery's energy requirements (Kumar et al. 2023).

Equation [4.3] can be used to calculate the battery's energy requirements.

$$Q_{Battery} = SOC + \int_0^t V_{bat} I_{bat} dt \qquad [4.3]$$

where SOC specifies the battery initial charge value, which can be assessed using equation [4.4].

$$B_{SOC} = \frac{Q}{Q_{bat.max}} \times 100\% \qquad [4.4]$$

4.3.4. *Power inverter*

The solar photovoltaic panel produces DC power. Because the load uses an AC source, the generated DC power must be converted into AC power. With built-in MPPT, an inverter converts DC electricity into AC power at a steady frequency. The inverter's rating determines how much electricity it can convert. Equation [4.5] can be used to get the inverter's rating (Ghasemi et al. 2013).

$$P_{inv}(t) = P_{pv}(t)\eta_{inv} \quad [4.5]$$

$$P_{pv}(t) = P_{sol}(t)N_{sol} \quad [4.6]$$

where the PV system's total output power is given by $P_{pv}(t)$ and η_{inv} is the inverter efficiency. P_{sol} (t) is the single solar PV cell output power, and N_{sol} gives the number of solar PV panels.

4.3.5. *Biomass gasifier*

Biogas gasification technology turns them into gaseous fuel, or producer gas, by partially burning the solid bio-residues under a controlled air supply. The combustion engine uses producer gas as an input fuel to generate energy. In the case of a biomass gasifier system, the capacity utilization factor determines the yearly output electricity E_{bg} of the rated biomass gasifier system P_{bg}, which may be computed as (Kushwah et al. 2022):

$$E_{bg} = P_{bg}(8{,}760 \times CUF) \quad [4.7]$$

The maximum size of the biomass gasifier system P_{bg}^{max} that may be installed in a particular location is mainly determined by the calorific value of biomass, availability, number of hours of use and system efficiency. The biomass gasifier's maximum rating can be computed as:

$$P_{bg}^{max} = \frac{M_{bg} \times 1{,}000 \times CV_{bm} \times \eta_{bg}}{365 \times 860 \times t_{bg}} \quad [4.8]$$

where M_{bg} is the entire biomass (Tonnes/year) available for the generation of power, η_{bg} gives the biomass gasifier system's efficiency per day and CV_{bm} is the calorific value of the biomass available.

4.4. Economic analyses

This section's objective is to confirm the model's economic viability. One of the main goals of the HOMER is the financial assessment, which determines the overall cost of the suggested system.

4.4.1. *LCOE*

LCOE is the ratio of the aggregate cost to the number of kWh generated across the project's entire lifespan. The LCOE of the selected model can be calculated using equation [4.9]:

$$LCOE = \left(\sum_{i=0}^{T}\left[\frac{C_i + L_i + O \text{ and } M_i + I_i}{(1+d)^i}\right]\right) / \left(\sum_{i=0}^{T}\left[\frac{E_i}{(1+d)^1}\right]\right) \quad [4.9]$$

The LCOE is an important metric that aids investors in choosing which hybrid energy system projects to fund.

4.4.2. *NPC*

The project lifecycle NPC is calculated by subtracting the current value of all revenue from the present value of the system's total cost. Equation [4.10] determines the NPC of each system component and its total cost. The minimum to NPC is the basis for the system selection in this study.

$$NPC = \frac{C_{ann,tot}}{CRF(i, R_{proj})} \quad [4.10]$$

$$CRF_{(i,N)} = i(1+i)^n/(1+i)^{(n-1)} \quad [4.11]$$

4.4.3. *Capacity factor*

The ratio of the system's actual energy production 24 hours a day for a year to its maximum output energy for the same period under rated power is known as the capacity factor. Typically, the capacity factor is shown as a percentage.

$$CF(\%) = \frac{E_{ac}}{P_0 \times 24 \times 365} \times 100 \quad [4.12]$$

4.4.4. *Resources estimation*

The coordinates for Aalborg are 9.9217° E longitude and 57.0488° N latitude. The average daily solar radiation in Aalborg ranges from 0.480 to 6.310 kWh/m^2/day, while the wind speed varies between 5.50 and 8.85 m/s. January had the highest wind speed in Aalborg, while June had the lowest. The location's temperature differs from 1.77°C to 16.8200°C, with the maximum temperature recorded in July and August, and the minimum temperature recorded in January and February. Figures 4.3–4.6 represent the variation of the solar irradiation, temperature, wind speed and biomass input. In this case study for analysis, biomass input is considered constant throughout the year and is taken as 8 tons per day. This is based on the local availability at Aalborg's well-established forestry waste supply chain that ensures stable feedstock (https://www.iea.org/reports/denmark-2023).

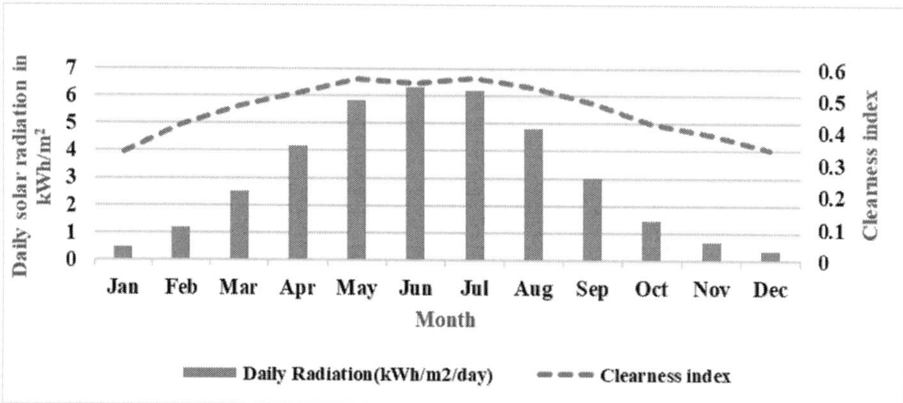

Figure 4.3. *Solar irradiation in kWh/m²*

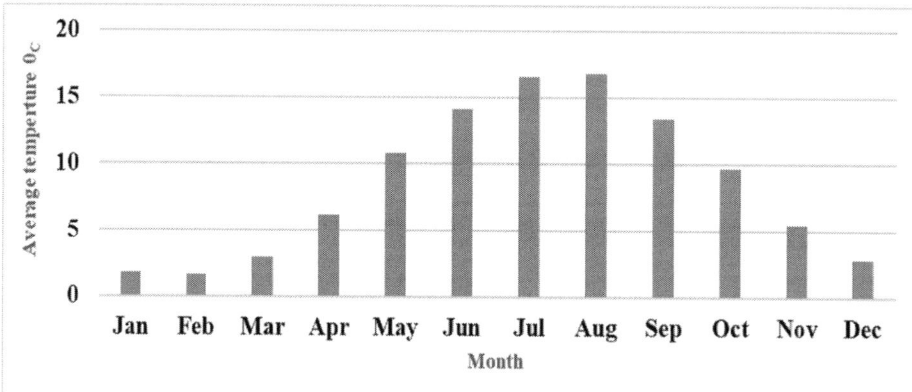

Figure 4.4. *Temperature variation*

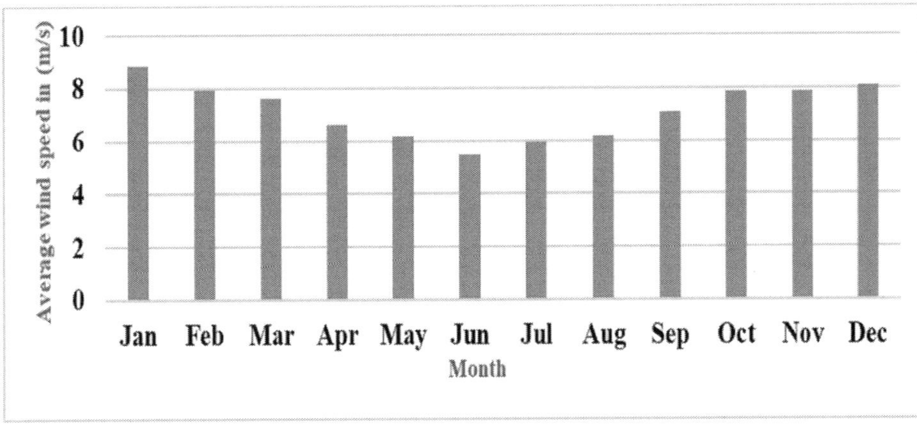

Figure 4.5. *Wind speed variation*

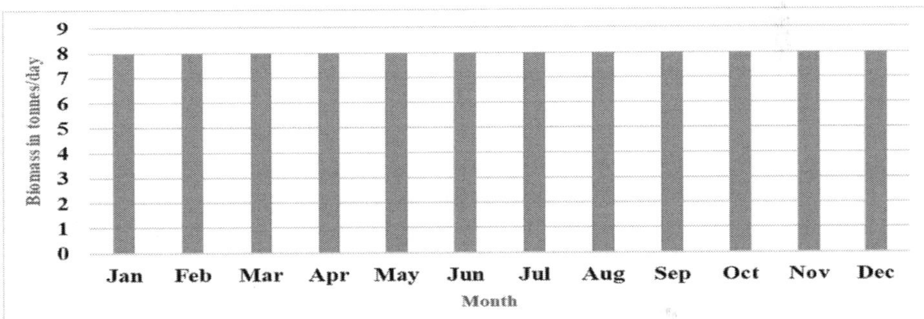

Figure 4.6. *Biomass input*

4.5. Results and discussion

The HOMER program was used to run the simulation. NASA provided RES data depending on the location entered into HOMER. The AC mini-grid's performance was evaluated, and estimates were made for the ideal RES, DG, battery and converter sizes. The AC mini-grid's techno-economic study is put into practice. A gated neighbourhood was taken into consideration while estimating the load variances. The load curve in Figure 4.7 illustrates the interpretation of the community load profile for a given day.

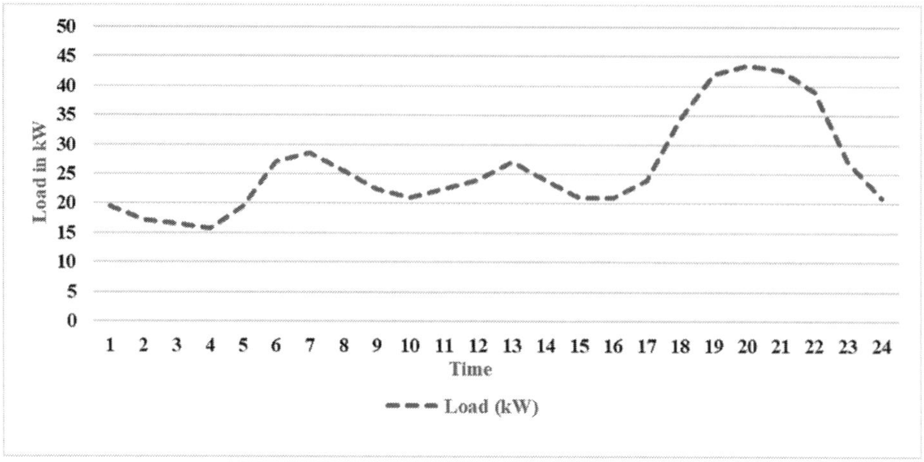

Figure 4.7. *Community load variation*

Figure 4.8 shows the flowchart of the proposed optimization algorithm executed in the HOMER software. The HES's primary architecture comprises battery storage, DG, biomass, wind and solar PV.

The initial optimization stage sets the main parameters, including the community load demand, solar irradiation, the chosen location's temperature, technical specifications and project economics.

The goal functions, such as NPC and LCOE, are calculated for each component size in the following step, which evaluates the component sizes.

Several simulation processes are carried out in each time step, depending on the main parameters and component sizes. The extra power generated by the solar PV, wind and biomass is stored in the battery bank if it exceeds the load requirement.

The battery's SOC will be checked if power generation falls short of load demand. If the battery's SOC is enough, it will provide the required energy.

After 8,760 h (1 year) of repetition, the best goal functions are obtained, and the NPC and LCOE of each configuration are confirmed.

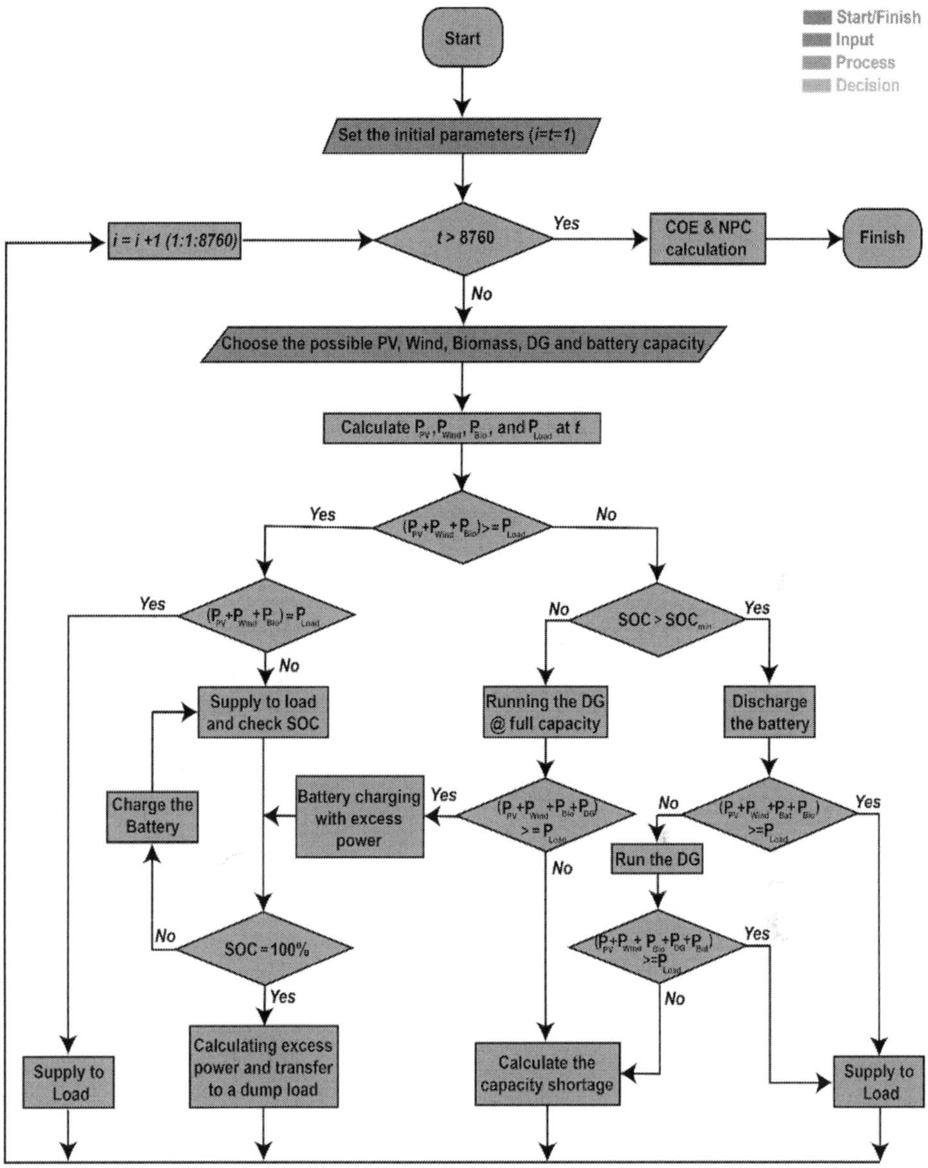

Figure 4.8. *Flowchart of the proposed work. For a color version of this figure, see www.iste.co.uk/sundaram/energysystems.zip*

4.5.1. *Different hybrid energy system configurations*

Figures 4.9 and 4.10 show different case studies used for the analysis. Each architecture consists of other energy sources that supply the load demand.

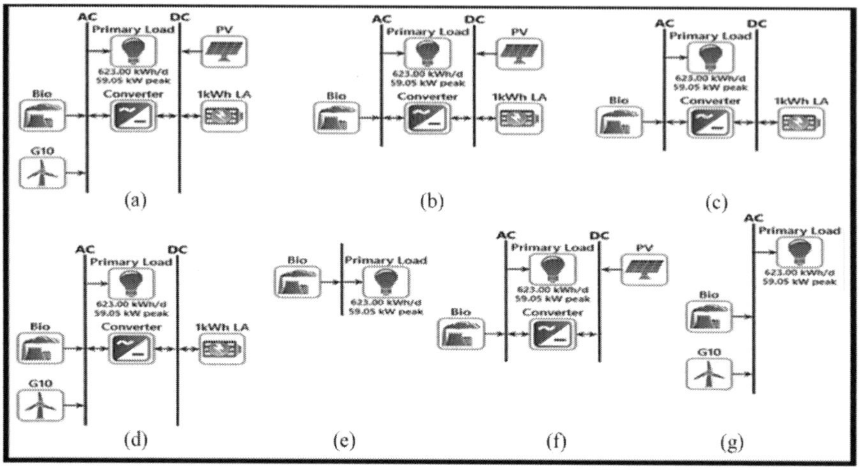

Figure 4.9. *Different hybrid energy system architecture: (a) PV/bio/wind/battery. (b) PV/bio/battery. (c) Bio/battery. (d) Bio/wind/battery. (e) Bio. (f) Bio/PV. (g) Bio/wind*

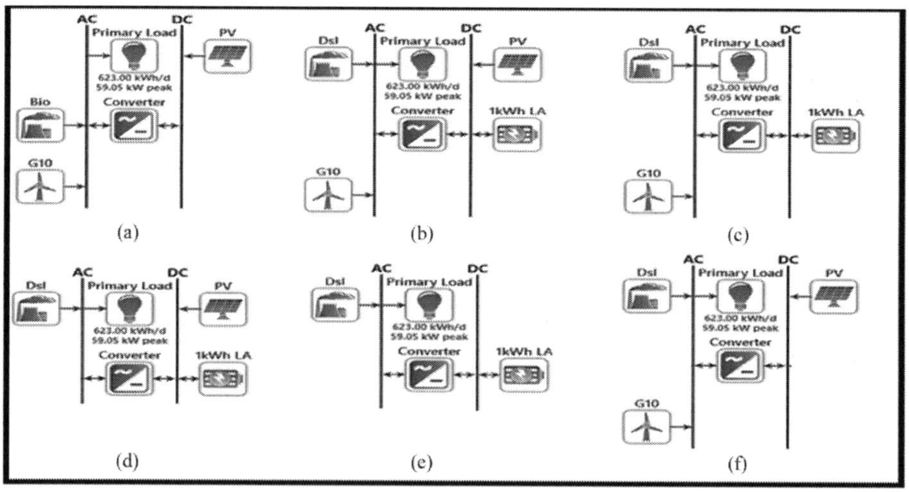

Figure 4.10. *(a) PV/bio/wind. (b) PV/DSI/wind/battery. (c) DSI/wind/battery. (d) PV/DSI/battery. (e) DSI/battery. (f) PV/DSI/wind*

4.5.2. *Case-1: PV/biomass/wind/battery are active*

The monthly energy provided by various RES and community load energy consumption is shown in Figure 4.11. The distribution of the overall energy generation was found to be as follows: 26% came from solar photovoltaics, 50.8% from biomass generators and 18.3% from wind turbines. The entire amount of energy generated by RES was 100%. In this instance, power from traditional energy sources was reduced while the whole load demand was satisfied by RES, such as solar PV, biomass and wind. Additionally, after fulfilling the total community load demand, the surplus energy generated is 8.56% (21,662 kWh/year). Assuming it could be efficiently used to connect dump loads or as reserves, this proportion of extra power was acceptable. It is well known that only a perfect system runs without generating surplus electricity, which this study did not.

Production	kWh/yr	%
Generic flat plate PV	65,539	26.0
Biomass Generator	140,824	55.8
Generic 10 kW	46,152	18.3
Total	252,515	100

Consumption	kWh/yr	%
AC Primary Load	227,343	100
DC Primary Load	0	0
Deferrable Load	0	0
Total	227,343	100

Quantity	kWh/yr	%
Excess Electricity	21,662	8.58
Unmet Electric Load	52.3	0.0230
Capacity Shortage	218	0.0957

Quantity	Value	Units
Renewable Fraction	100	%
Max. Renew. Penetration	391	%

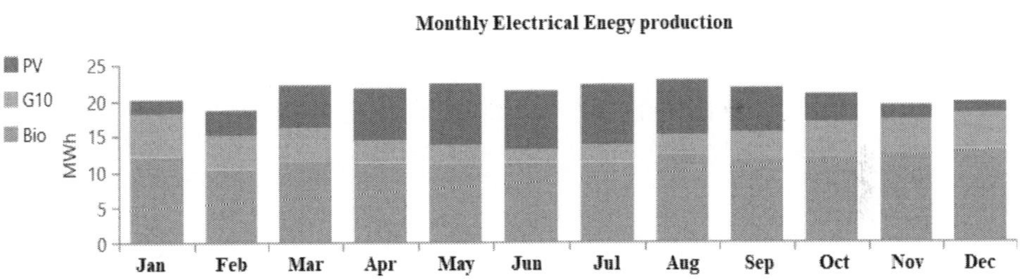

Figure 4.11. *Monthly electrical energy production, annual electrical energy production and consumption. For a color version of this figure, see www.iste.co.uk/sundaram/energysystems.zip*

According to the findings, the annual unmet electrical load was predicted to be 0.0230% (52.3 kWh/year), while the capacity shortfall of the suggested system was only 0.0957%(218 kWh/year). The projected annual CO_2 emissions were 42.2 kg. An estimated 673 tons of fuel were needed annually. The specific fuel consumption is assessed as 3.34 kg/kWh. The variation of the biomass generator is shown in Figure 4.12. It was observed that the energy produced by the biomass generator is maximum from 18:00 to 22:00 h, which is the peak load hours. During peak hours, it was discovered that the biomass generator was running at full capacity. Figures 4.13–4.15 illustrate the solar PV, wind turbine power output and the SOC of the battery.

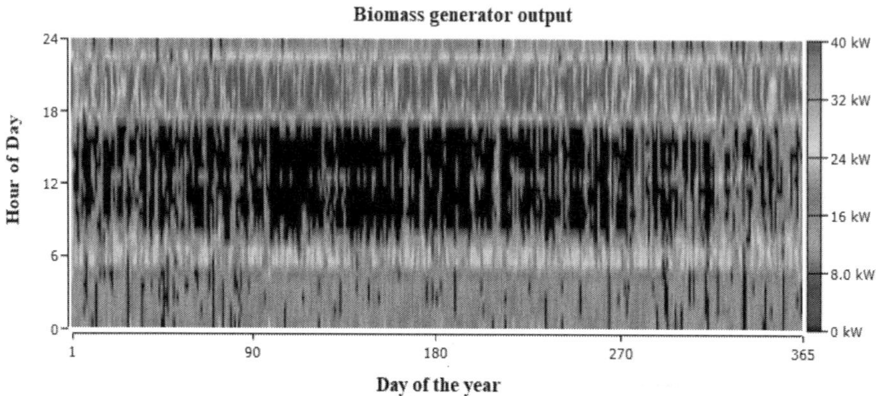

Figure 4.12. *Biomass generator output. For a color version of this figure, see www.iste.co.uk/sundaram/energysystems.zip*

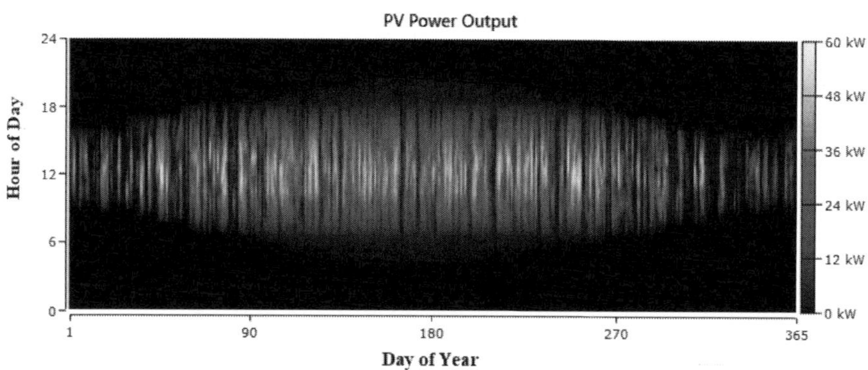

Figure 4.13. *Solar PV output. For a color version of this figure, see www.iste.co.uk/sundaram/energysystems.zip*

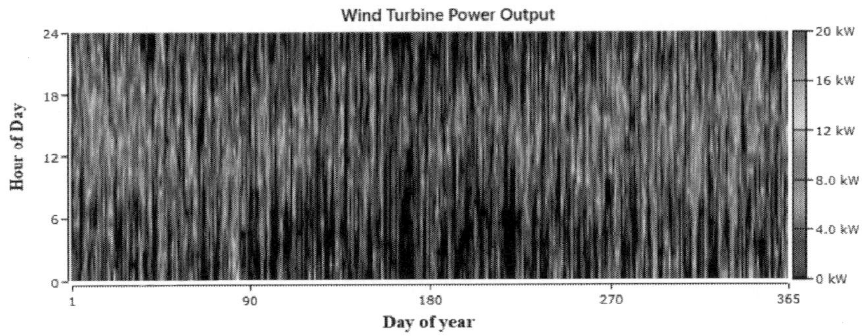

Figure 4.14. *Wind generator output. For a color version of this figure, see www.iste.co.uk/sundaram/energysystems.zip*

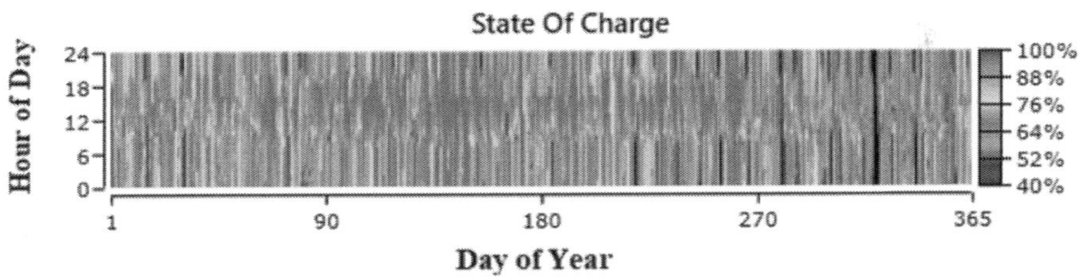

Figure 4.15. *SOC of the battery. For a color version of this figure, see www.iste.co.uk/sundaram/energysystems.zip*

Table 4.1 displays the results of assessing each source's capacity factor, minimum and maximum electrical outputs, annual operation in hours and ideal RES size. The table shows that the wind turbine ran for 7,468 h annually, the maximum amount of time it could. The wind turbine was estimated to use 26.3% of its capacity. Every day, the biomass generator ran for 6,358 h. It was estimated that the biomass generator's capacity factor was 40.2%.

The optimum battery size for the proposed community load at Aalborg was estimated to be 74 kWh. Figure 4.15 shows the SOC of the battery. The energy input of the battery per annum was estimated to be 6,392 kWh/year. Note that 5,140 kWh of energy were produced annually, while 28.6 kWh of battery storage was depleted overall. The battery storage's overall loss was calculated to be 1,282 kWh annually. The battery system's projected yearly throughput was 5,746 kWh.

PV/biomass/wind and battery storage					
Components	**Size**	**Annual hours of operation (h)**	**Minimum electrical output**	**Maximum electrical output**	**Capacity factor in %**
Solar PV	60 kW	4,370	0	58.7 kW	12.5
Wind turbine	20 kW	7,468	0	20 kW	26.3
Biomass	40 kW	6,358	12 kW	40 kW	40.2
Battery storage	74 kWh	–	–	–	–
Converter	29.5 kW	4,719	–	–	–

Table 4.1. *Technical characteristics of the optimal system*

4.5.3. *Case 2: when solar PV/biomass/batteries are active*

The monthly energy produced by various RES and battery storage is displayed in Figure 4.16. The communal load's energy use is shown. It was observed that the total energy generation was dispersed as 27% from solar PV and 73% from biomass generators. Renewable sources provided 100% of the energy. In this instance, power from traditional energy sources is reduced by the total load demand satisfied by the RES, such as solar PV and biomass. After fulfilling the community's overall load demand, the extra energy generated comes to 4.87% (11,819 kWh/year). The findings demonstrated no annual unmet electrical load or capacity limitation in the suggested system. The anticipated yearly CO_2 emissions were 55.6 kg. An estimated 882 tons of fuel were needed annually. It is estimated that the specific fuel consumption is 3.49 kg/kWh.

The variation of the biomass generator is shown in Figure 4.17. It was observed that the energy produced by the biomass generator is maximum from 18:00 to 22:00 h, which is the peak load hours. The biomass generator was found to be operating at its total capacity during peak hours. Figure 4.18 indicates the output of the solar PV.

Table 4.2 displays the results of assessing each source's capacity factor, minimum and maximum electrical outputs, annual operation in hours and ideal RES size. The table shows that the biomass generator ran for 7,025 h a year, the maximum time possible. The wind turbine's capacity utilization was calculated to be 40.4%.

It was determined that a battery of 63 kWh would be the ideal size for the suggested community load. Figure 4.19 shows the battery's SOC. The annual energy input was calculated to be 3,899 kWh. The anticipated total battery storage depletion was 29.2 kWh/year, whereas the energy output was 3145 kWh/year. It was calculated that the battery storage will lose 783 kWh annually. The battery system's yearly throughput was predicted to be 3,517 kWh.

Production	kWh/yr	%
Generic flat plate PV	65,539	27.0
Biomass Generator	177,101	73.0
Total	242,640	100

Consumption	kWh/yr	%
AC Primary Load	227,395	100
DC Primary Load	0	0
Deferrable Load	0	0
Total	227,395	100

Quantity	kWh/yr	%
Excess Electricity	11,819	4.87
Unmet Electric Load	0	0
Capacity Shortage	0	0

Quantity	Value	Units
Renewable Fraction	100	%
Max. Renew. Penetration	292	%

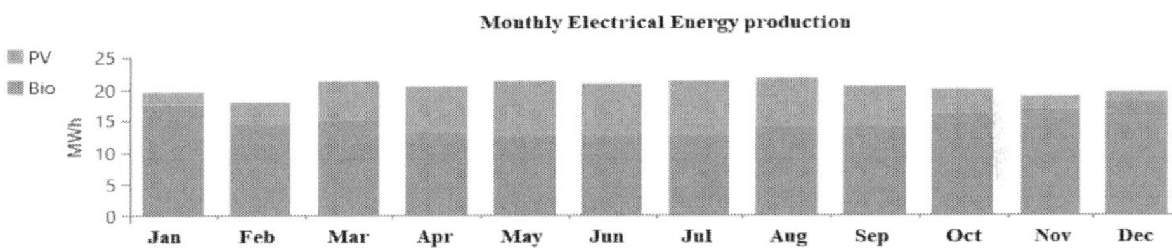

Figure 4.16. *Monthly electrical energy production, annual electrical energy production and consumption. For a color version of this figure, see www.iste.co.uk/sundaram/energysystems.zip*

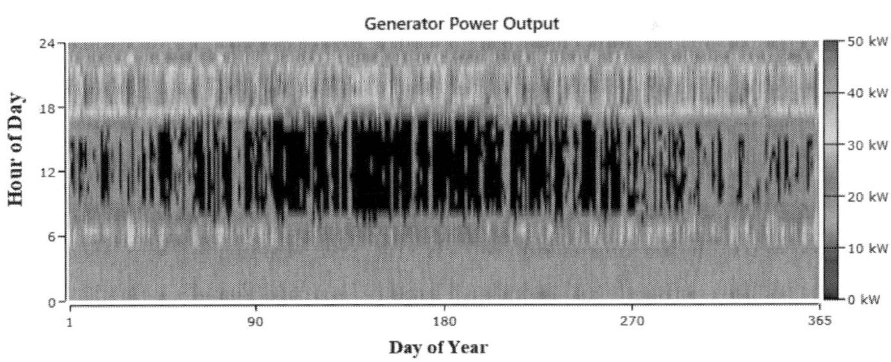

Figure 4.17. *Biomass generator output. For a color version of this figure, see www.iste.co.uk/sundaram/energysystems.zip*

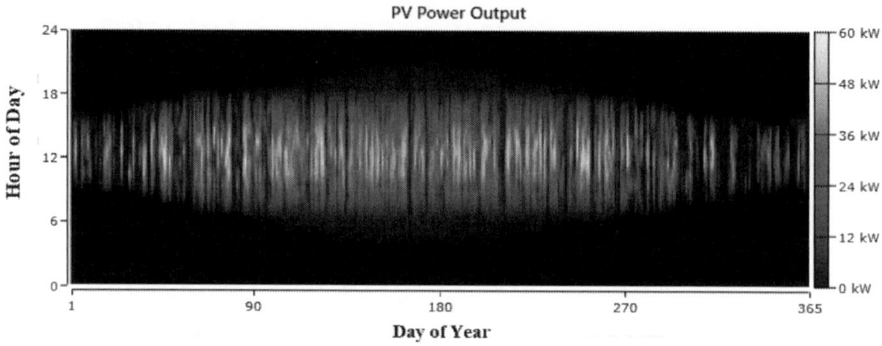

Figure 4.18. *Solar PV output. For a color version of this figure, see www.iste.co.uk/sundaram/energysystems.zip*

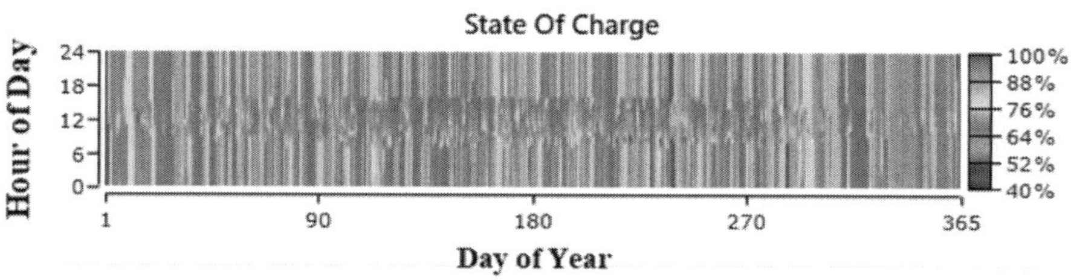

Figure 4.19. *SOC of the battery. For a color version of this figure, see www.iste.co.uk/sundaram/energysystems.zip*

PV/biomass/battery storage					
Components	**Size**	**Annual hours of operation (h)**	**Minimum electrical output**	**Maximum electrical output**	**Capacity factor in %**
Solar PV	60kW	4,370	0	58.7 kW	12.5
Biomass	50 kW	7,025	15 kW	50 kW	40.4
Battery storage	63 kWh	–	–	–	–
Converter	37.4 kW	–	–	–	–

Table 4.2. *Technical specification of the optimal system*

4.5.4. *Case 3: when biomass alone is active*

Figure 4.20 shows the monthly energy generated by various RES. It was observed that the total energy generation from the biomass generator is 100% of the total load demands, and 100% of the energy was generated from renewable sources. In this case, a biomass generator alone met the entire load demand while curtailing power from conventional energy sources. The extra energy generated after satisfying the community's overall load demand is 0.0594% (135 kWh/year). The findings verified no annual unmet electrical load or capacity limitation in the suggested system. The projected CO_2 emissions per year were 70.5 kg. An estimated 1,118 tons of fuel per year were needed. The specific fuel consumption is calculated as 3.44 kg/kWh. The variation of the biomass generator is shown in Figure 4.21. The biomass generator is operated throughout the day. Still, the maximum power generated and utilized was during the peak demand between 6 pm. and 10 pm.

Production	kWh/yr	%
Biomass Generator	227,535	100
Total	227,535	100

Consumption	kWh/yr	%
AC Primary Load	227,372	100
DC Primary Load	0	0
Deferrable Load	0	0
Total	227,372	100

Quantity	kWh/yr	%
Excess Electricity	135	0.0594
Unmet Electric Load	22.6	0.00990
Capacity Shortage	208	0.0914

Quantity	Value	Units
Renewable Fraction	100	%
Max. Renew. Penetration	0	%

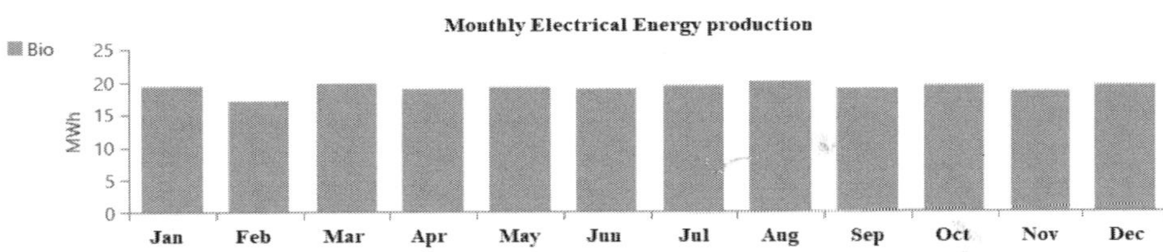

Figure 4.20. *Monthly electrical energy production*

Table 4.3 displays the results of assessing each source's capacity factor, minimum and maximum electrical outputs, annual operation in hours and ideal RES size. The table indicates that the biomass produced ran for 8,760 h annually. The optimal battery size for the proposed community load was 21 kWh. From Figure 4.22, it was seen that the battery SOC is always at its maximum, and the battery usage is significantly less compared to case 1 and case 2. It was estimated that the annual energy input was 131 kWh. An estimated 116 kWh of energy was produced annually, while 12.6 kWh of battery storage was depleted overall. An estimated 27.5 kWh of battery storage was lost annually. The battery system's projected yearly throughput was 130 kWh.

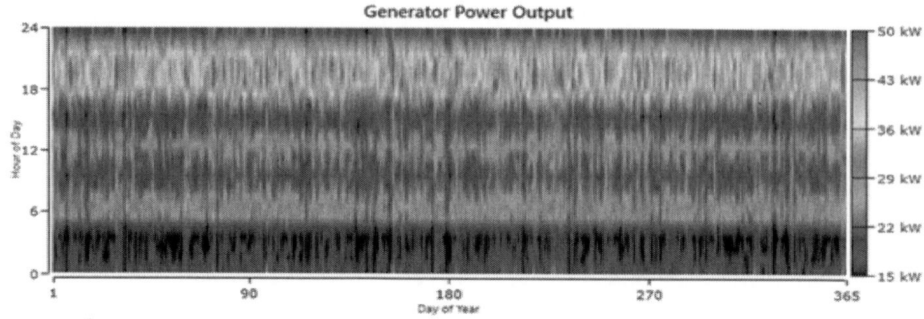

Figure 4.21. *Biomass generator output. For a color version of this figure, see www.iste.co.uk/sundaram/energysystems.zip*

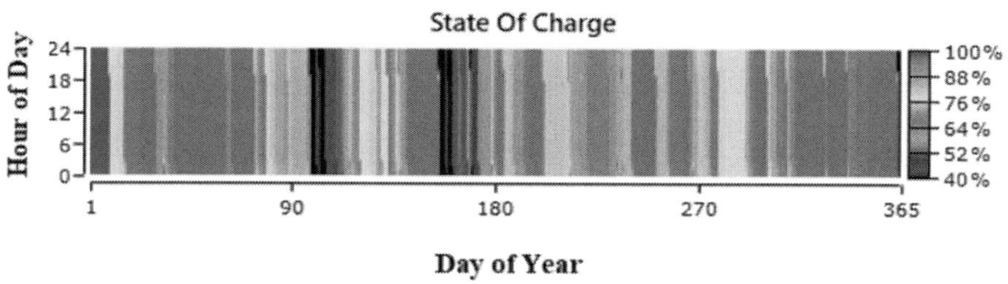

Figure 4.22. *SOC of the battery. For a color version of this figure, see www.iste.co.uk/sundaram/energysystems.zip*

Biomass and battery					
Components	**Size**	**Annual hours of operation (h)**	**Minimum electrical output**	**Maximum electrical output**	**Capacity factor in %**
Biomass	50 kW	8,760	15 kW	50 kW	51.9
Battery storage	21 kWh	–	–	–	–
Converter	6.49 kW	–	–	5.21 KW	–

Table 4.3. *Technical specification of the optimal system*

4.5.5. *Case 4: when biomass and wind generators are active*

Production	kWh/yr	%
Biomass Generator	146,961	61.4
Generic 10 kW	92,304	38.6
Total	239,265	100

Consumption	kWh/yr	%
AC Primary Load	227,322	100
DC Primary Load	0	0
Deferrable Load	0	0
Total	227,322	100

Quantity	kWh/yr	%
Excess Electricity	10,190	4.26
Unmet Electric Load	73.1	0.0321
Capacity Shortage	202	0.0890

Quantity	Value	Units
Renewable Fraction	100	%
Max. Renew. Penetration	290	%

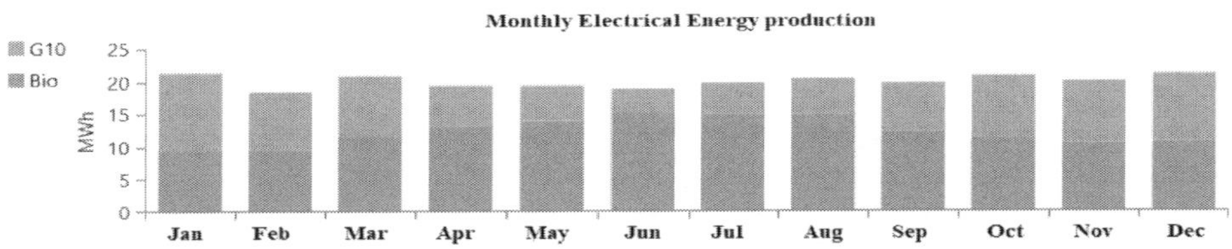

Figure 4.23. *Monthly electrical energy production and annual electrical energy production and consumption. For a color version of this figure, see www.iste.co.uk/sundaram/energysystems.zip*

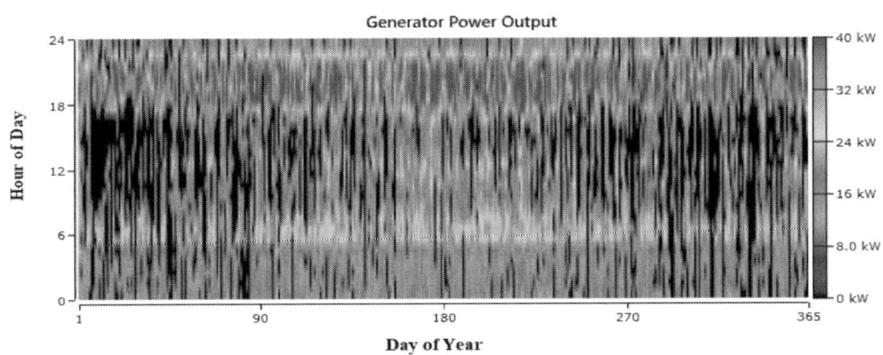

Figure 4.24. *Biomass generator output. For a color version of this figure, see www.iste.co.uk/sundaram/energysystems.zip*

The monthly energy provided by various RES is shown in Figure 4.23. The communal load's energy use is also shown. It was found that the energy generated by a wind turbine accounts for 35.6% of the overall demand. In comparison, the energy generated by a biomass generator accounts for 61.4% of the total demand. The entire amount of energy generated by RES was 100%. In this case, the RES meets the whole load. Biomass and wind generators are used while curtailing power from conventional energy sources. After satisfying the entire load demand of the neighborhood, the excess energy generated amounts to 4.26% (10,190 kWh/year). According to the findings, the annual unmet electrical load was 0.0321%, and the capacity shortfall of the suggested system was also 0.08%. The projected CO_2 emissions per year were 44.2 kg. The projected annual fuel requirement was 701 tons. The specific fuel consumption is estimated as 3.34 kg/kWh. The variation of the biomass generator is shown in Figure 4.24. It was noted that the energy produced by the biomass generator is maximum from 18:00 to 22:00 h, which is the peak load hours. The biomass generator was found to be operating at its total capacity during peak hours. Figure 4.25 shows the output power variation of the wind turbine.

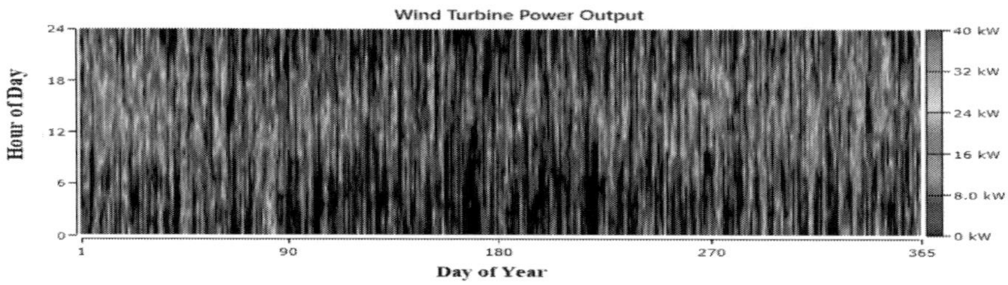

Figure 4.25. *Wind turbine power output. For a color version of this figure, see www.iste.co.uk/sundaram/energysystems.zip*

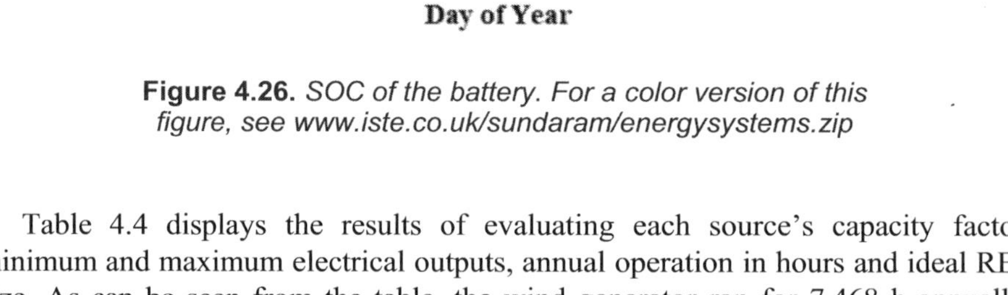

Figure 4.26. *SOC of the battery. For a color version of this figure, see www.iste.co.uk/sundaram/energysystems.zip*

Table 4.4 displays the results of evaluating each source's capacity factor, minimum and maximum electrical outputs, annual operation in hours and ideal RES size. As can be seen from the table, the wind generator ran for 7,468 h annually, while the biomass generator ran for 6,604 h, the highest amount of time possible.

The optimum size of the battery for the proposed community load was assessed as 131 kWh. Figure 4.24 illustrates the SOC of the battery. It was estimated that the annual energy input was 5,997 kWh. The overall battery storage depletion was assessed to be 3.81 kWh/year, whereas the energy output was predicted to be 4,801 kWh/year. An estimated 1,200 kWh of battery storage was lost annually. The battery system's projected yearly throughput was 5,367 kWh.

Technical characteristics of the optimal system					
Biomass/wind and battery					
Components	**Size**	**Annual hours of operation (h)**	**Minimum electrical output**	**Maximum electrical output**	**Capacity factor in %**
Biomass	40 kW	6,604	12 kW	40 kW	41.9
Wind	40 kW	7,468	0	40 kW	26.3
Battery storage	131 kWh	–	–	–	–
Converter	26.4 kW	–	–	–	–

Table 4.4. *Technical specification of the optimal system*

4.5.6. *Case 5: when solar PV, DG and wind generators are active*

The monthly energy produced by various RES and batteries in Aalborg is displayed in the Figure. 4.27. The communal load's energy use is also shown. It was found that the energy generated by wind turbines (49.6% of the total load demand), diesel generators (26.7%) and solar photovoltaics (23.5%) accounted for the majority of the energy produced. The total energy generated by RES accounts for 67%. In this case, only 67% of the total load demand is met by the RES. Solar PV and wind generators curtail power from conventional energy resources. When the entire load demand of the community is met, the excess energy produced comes to 13.9% (38,920 kWh/year). The findings indicated that the capacity shortfall of the suggested system was 0.0884%, and the annual unmet electrical load was 0.0167%.

The anticipated annual CO_2 emissions were 75,292 kg. The expected amount of gasoline used was 28,594 L/year. The specific fuel consumption is estimated as 0.381 L/kWh.

The variation of DG output is shown in Figure 4.28. It was observed that the maximum energy produced by the DG during the peak load hours is from 18:00 to 22:00 h. The DG was observed to be functioning at its full capacity during peak hours. Figure 4.29 indicates the variation of the wind turbine output.

Production	kWh/yr	%
Generic flat plate PV	65,539	23.5
Diesel Generator	75,117	26.9
Generic 10 kW	138,456	49.6
Total	279,112	100

Consumption	kWh/yr	%
AC Primary Load	227,357	100
DC Primary Load	0	0
Deferrable Load	0	0
Total	227,357	100

Quantity	kWh/yr	%
Excess Electricity	38,920	13.9
Unmet Electric Load	38.1	0.0167
Capacity Shortage	201	0.0884

Quantity	Value	Units
Renewable Fraction	67.0	%
Max. Renew. Penetration	628	%

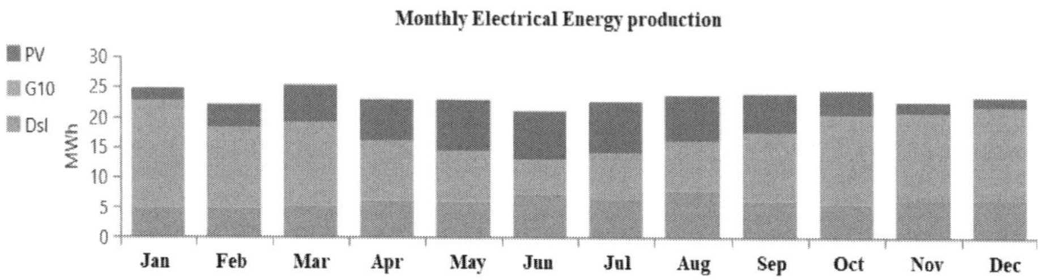

Figure 4.27. *Monthly energy produced, annual energy produced and consumed. For a color version of this figure, see www.iste.co.uk/sundaram/energysystems.zip*

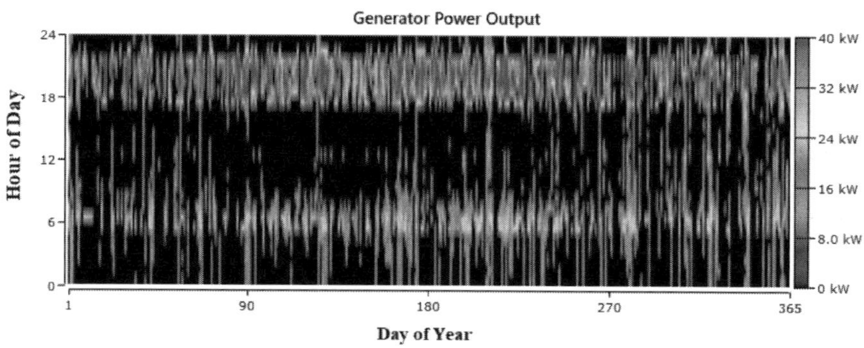

Figure 4.28. *Diesel generator output power. For a color version of this figure, see www.iste.co.uk/sundaram/energysystems.zip*

Table 4.5 displays the results of assessing each source's capacity factor, minimum and maximum electrical outputs, annual operation in hours and ideal RES size. The table shows that the DG ran for 3,067 h annually, the wind generator ran for 7,468 h annually and the solar PV installed for 4,070 h annually.

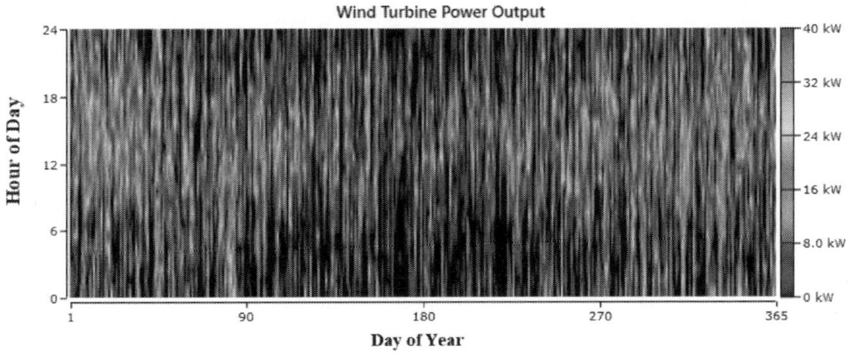

Figure 4.29. *Wind turbine output power. For a color version of this figure, see www.iste.co.uk/sundaram/energysystems.zip*

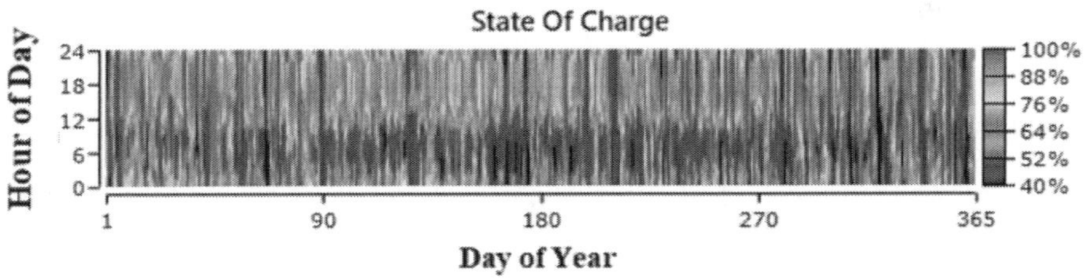

Figure 4.30. *SOC of the battery. For a color version of this figure, see www.iste.co.uk/sundaram/energysystems.zip*

PV/biomass/wind/DsI/battery					
Components	**Size**	**Annual hours of operation (h)**	**Minimum electrical output**	**Maximum electrical output**	**Capacity factor in %**
PV	60 kW	4,370	0	58.7	12.5
DsI	40 kW	3,067	12 kW	40 kW	21.4
Wind	60 kW	7,468	0	60 kW	26.3
Battery storage	312 kWh	–	–	–	–
Converter	30.6 kW	4,942	–	–	–

Table 4.5. *Technical specification of the optimal system*

Note that 312 kWh was the ideal battery size for the suggested load. The battery’s SOC is shown in Figure 4.30. It was estimated that the annual energy input was 44,394 kWh. An estimated 35,618 kWh of energy were produced annually, and 115 kWh of battery storage were depleted overall. An estimated 8,891 kWh of battery storage was lost annually. The battery system's projected yearly throughput was 39,822 kWh.

4.5.7. *Case 6: when the diesel generator and the wind generators are active*

The monthly energy produced by various RES is displayed in Figure 4.31, as well as the community load's yearly energy consumption. It was found that the energy generated by wind turbines accounts for 54.4% of the total load demands, while the energy generated by diesel generators accounts for 45.4%. Note that 49.4% of the total energy generated by RES came from this source. In this instance, wind generators are utilized while reducing electricity from traditional energy sources, and the RES barely meets 49.4% of the total load requirement. When the entire load demand of the community is met, the excess energy produced comes to 6.89% (17,471 kWh/year). The findings indicated that the capacity shortfall of the suggested system was 0.00370%, and the annual unmet electrical load was 0%. The anticipated yearly CO_2 emissions were 126,227 kg. It was calculated that 47,934 L of fuel will be needed annually. It is estimated that the specific fuel consumption is 0.417 L/kWh. Figures 4.32 and 4.33 illustrate the fluctuation in wind turbine and DG output. The peak load hours, from 18:00 to 22:00 h, were found to be when the DG produced the most energy. It was discovered that during peak hours, the DG was running at maximum efficiency.

Production	kWh/yr	%
Diesel Generator	115,002	45.4
Generic 10 kW	138,456	54.6
Total	253,457	100

Consumption	kWh/yr	%
AC Primary Load	227,395	100
DC Primary Load	0	0
Deferrable Load	0	0
Total	227,395	100

Quantity	kWh/yr	%
Excess Electricity	17,471	6.89
Unmet Electric Load	0	0
Capacity Shortage	8.52	0.00370

Quantity	Value	Units
Renewable Fraction	49.4	%
Max. Renew. Penetration	434	%

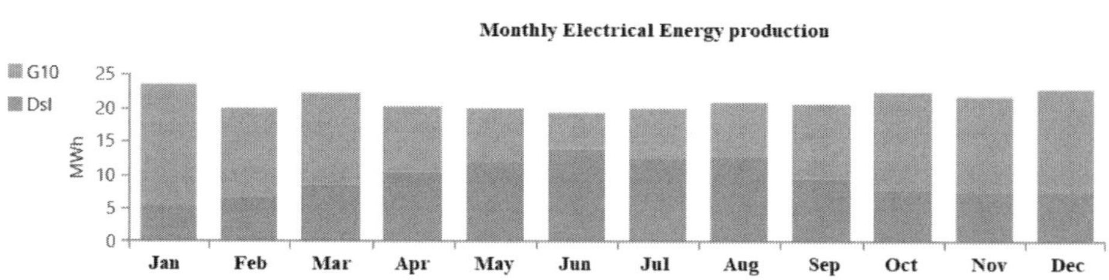

Figure 4.31. *Monthly electrical energy produced and annual energy produced and consumed. For a color version of this figure, see www.iste.co.uk/sundaram/energysystems.zip*

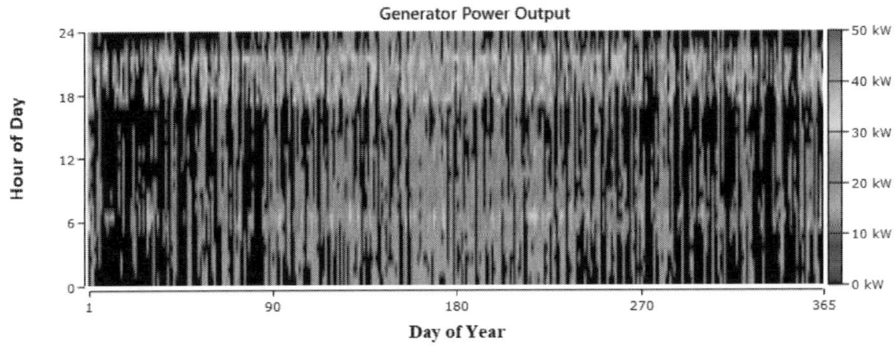

Figure 4.32. *DG output power. For a color version of this figure, see www.iste.co.uk/sundaram/energysystems.zip*

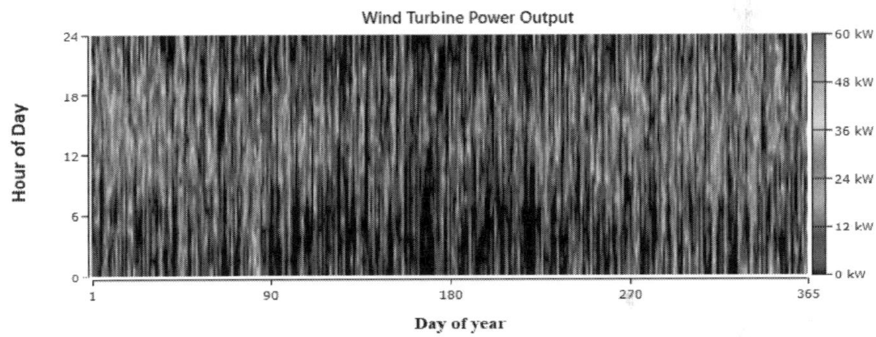

Figure 4.33. *Wind turbine output power. For a color version of this figure, see www.iste.co.uk/sundaram/energysystems.zip*

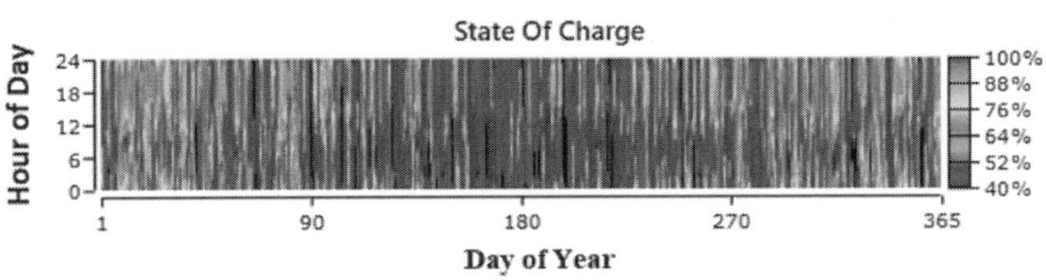

Figure 4.34. *SOC of the battery. For a color version of this figure, see www.iste.co.uk/sundaram/energysystems.zip*

Table 4.6 displays the results of assessing each source's capacity factor, minimum and maximum electrical outputs, annual operation in hours and ideal RES size. The table shows that the wind generator ran for 7,468 h annually, while the DG ran for 4,736 h, the maximum amount of time it can function.

It was determined that 275 kWh was the ideal battery size for the suggested community load. Figure 4.34 displays the battery's SOC. It was estimated that the annual energy input was 29,640 kWh. The overall battery storage depletion was predicted to be 96.8 kWh/year, whereas the energy output was projected to be 23798 kWh/year. An estimated 5,938 kWh of battery storage was lost annually. The battery system's projected yearly throughput was 26,607 kWh.

DG/wind and battery					
Components	**Size**	**Annual hours of operation (h)**	**Minimum electrical output**	**Maximum electrical output**	**Capacity factor in %**
DG	50 kW	4,736	15 kW	50 kW	26.3
Wind	60 kW	7,468	0	60 kW	26.3
Battery storage	275 kWh	–	–	–	–
Converter	28.7 kW	1,797	–	–	–

Table 4.6. *Technical specification of the optimal system*

4.5.8. *Case 7: when solar PV and DG are active*

The monthly energy produced by various RES is displayed in Figure 4.35. The communal load's energy use is also shown. It was found that the energy generated by solar PV accounts for 27.2% of the overall load demand, whereas the energy generated by diesel generators accounts for 72.8% of the load demand. In all, RES generated 22.9% of the energy. In this instance, solar PV was utilized while reducing electricity from traditional energy sources, and the RES only satisfied 22.9% of the entire load demand. The extra energy generated after satisfying the community's overall load demand is 3.87% (9,317 kWh/year). According to the findings, the annual unmet electrical load was 0%, and the capacity shortfall of the suggested system was likewise 0.00160%. The anticipated yearly CO_2 emissions were 118,164 kg. Note that 71,455 L of fuel per year was predicted to be needed. It is estimated that the specific fuel consumption is 0.408 L/kWh. The DG output variation is displayed in Figure 4.36. The peak load hours, from 18:00 to 22:00 h, were found to be when the DG produced the most energy. It was discovered that the

DG was running at full capacity during peak hours. The volatility of the solar PV production is shown in Figure 4.37.

Production	kWh/yr	%
Generic flat plate PV	65,539	27.2
Diesel Generator	175,307	72.8
Total	240,846	100

Consumption	kWh/yr	%
AC Primary Load	227,395	100
DC Primary Load	0	0
Deferrable Load	0	0
Total	227,395	100

Quantity	kWh/yr	%
Excess Electricity	9,317	3.87
Unmet Electric Load	0	0
Capacity Shortage	3.58	0.00160

Quantity	Value	Units
Renewable Fraction	22.9	%
Max. Renew. Penetration	292	%

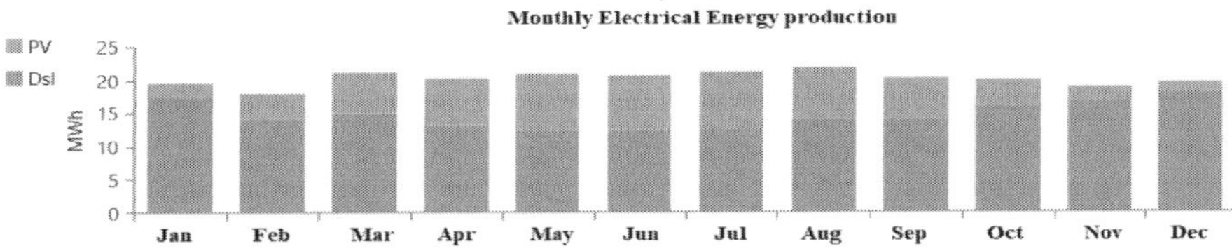

Figure 4.35. *Monthly electrical energy production and annual energy production and consumption. For a color version of this figure, see www.iste.co.uk/sundaram/energysystems.zip*

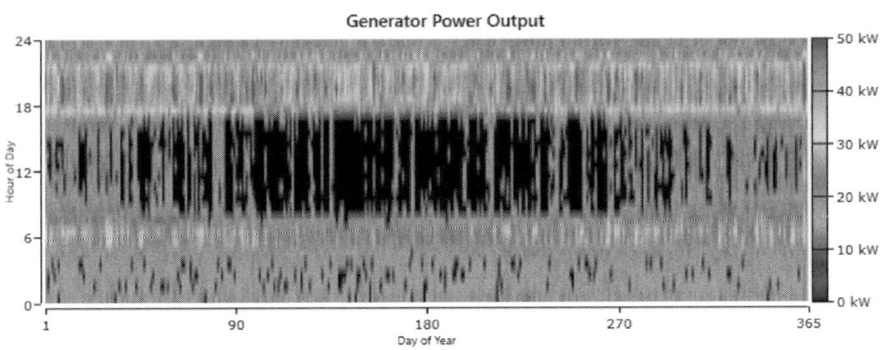

Figure 4.36. *DG output power. For a color version of this figure, see www.iste.co.uk/sundaram/energysystems.zip*

Table 4.7 displays the results of assessing each source's capacity factor, minimum and maximum electrical outputs, annual operation in hours and ideal RES size. The chart indicates that the solar PV functioned for 4,370 h annually, whereas the DG operated for the most significant amount of hours, 6,907 h annually.

It was determined that 70 kWh was the ideal battery size for the suggested community load. Figure 4.37 displays the battery's SOC. It was estimated that the annual energy intake was 7,017 kWh. An estimated 5,645 kWh of energy was produced annually, while 35.1 kWh of battery storage was depleted overall. An estimated 1,407 kWh of battery storage was lost annually. The battery system's projected yearly throughput was 6,311 kWh.

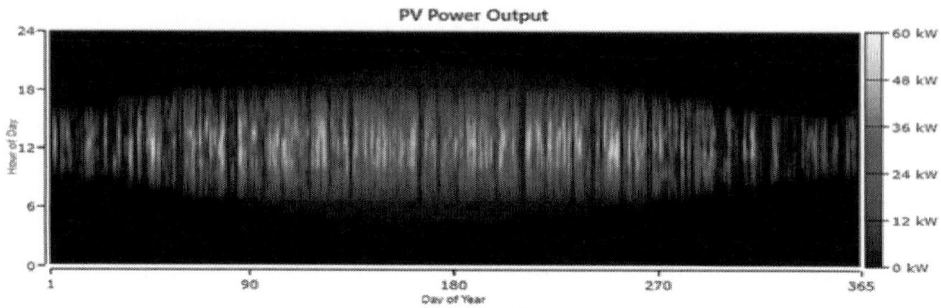

Figure 4.37. *Solar PV output. For a color version of this figure, see www.iste.co.uk/sundaram/energysystems.zip*

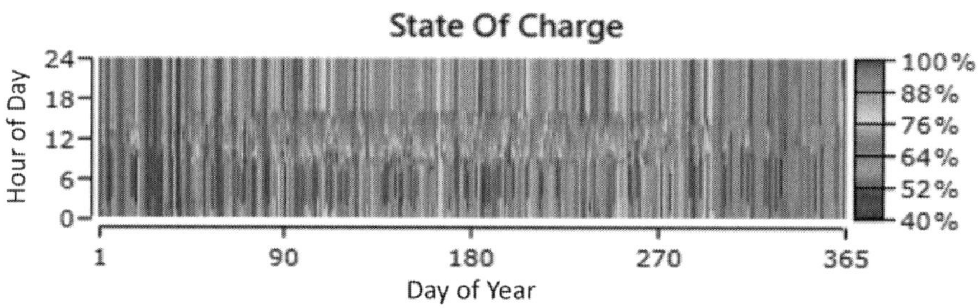

Figure 4.38. *SOC of the battery. For a color version of this figure, see www.iste.co.uk/sundaram/energysystems.zip*

DSI/solar and battery					
Components	**Size**	**Annual hours of operation (h)**	**Minimum electrical output**	**Maximum electrical output**	**Capacity factor in %**
DsI	50 kW	6,907	15 kW	50 kW	40.4
Solar PV	60 kW	4,370	0	60 kW	12.5
Battery storage	70 kWh	–	–	–	–
Converter	37.5 kW	1,797	–	–	–

Table 4.7. *Technical specification of the optimal system*

4.5.9. *Case 8: when a diesel generator alone supplies the load demand*

Figure 4.39 shows the energy consumed by the community load and the energy produced by the diesel generator. It was observed that the total load demand is met by the DG alone in this case. The extra energy generated after satisfying the community's overall load demand is 0.0594% (135 kWh/year).

Production	kWh/yr	%
Diesel Generator	227,535	100
Total	227,535	100

Consumption	kWh/yr	%
AC Primary Load	227,372	100
DC Primary Load	0	0
Deferrable Load	0	0
Total	227,372	100

Quantity	kWh/yr	%
Excess Electricity	135	0.0594
Unmet Electric Load	22.6	0.00990
Capacity Shortage	208	0.0914

Quantity	Value	Units
Renewable Fraction	0	%
Max. Renew. Penetration	0	%

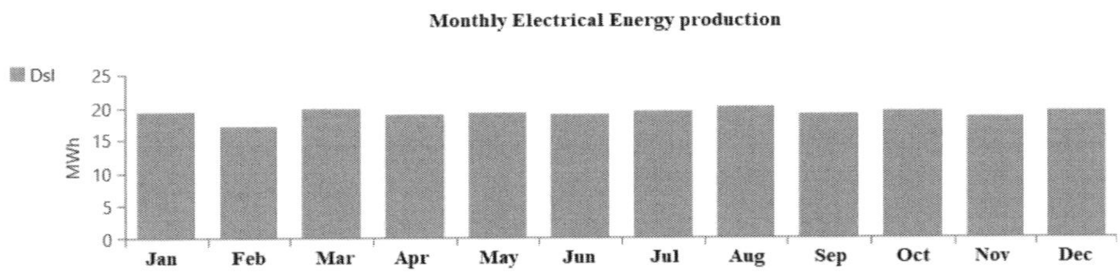

Figure 4.39. *Monthly electrical energy production and annual energy produced and consumed*

According to the findings, the annual unmet electrical load was 0%, and the capacity shortfall of the suggested system was likewise 0.0099%. The anticipated yearly CO_2 emissions were 242,066 kg. The expected amount of gasoline used was 91,924 L annually. It is estimated that the specific fuel consumption is 0.404 L/kWh. The DG output variation is displayed in Figure 4.40. The peak load hours, from 18:00 to 22:00 h, were found to be when the DG produced the most energy. During peak hours, the DG was running at maximum capacity; during off-peak load hours, it was running at a lower rating.

Table 4.8 displays the results of assessing the system's ideal size, annual operation in hours, minimum and maximum electrical output and a capacity factor for each source. The data show that the DG ran for 8,760 h annually, which is the maximum time it could operate.

It was determined that 21 kWh was the ideal battery size for the suggested community load. It was observed from Figure 4.41 that the battery SOC is at its highest point during the day. It was estimated that the annual energy input was

131 kWh. An estimated 116 kWh of energy was produced annually, while 12.6 kWh of battery storage was depleted overall. An estimated 27.5 kWh of battery storage was lost annually. The battery system's projected yearly throughput was 130 kWh.

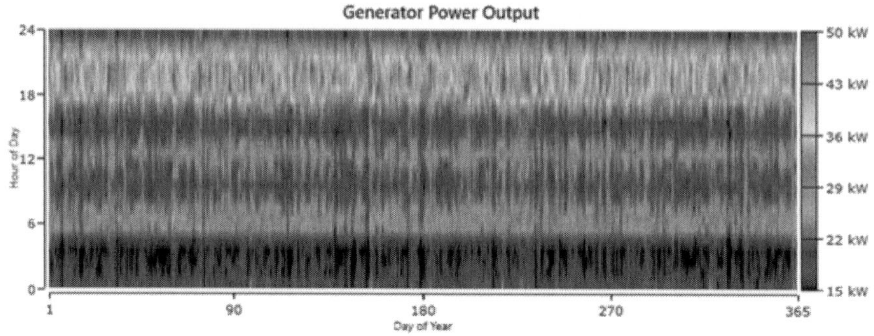

Figure 4.40. *DG power output. For a color version of this figure, see www.iste.co.uk/sundaram/energysystems.zip*

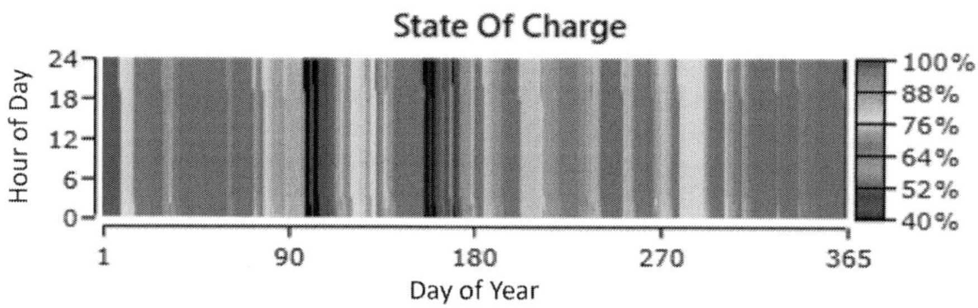

Figure 4.41. *SOC of the battery. For a color version of this figure, see www.iste.co.uk/sundaram/energysystems.zip*

DSI/solar and battery					
Components	**Size**	**Annual hours of operation (h)**	**Minimum electrical output**	**Maximum electrical output**	**Capacity factor in %**
DG	50 kW	8,760	15 kW	50 kW	51.9
Battery storage	21 kWh	–	–	–	–
Converter	6.49 kW	1,797	–	–	–

Table 4.8. *Technical characteristics of the optimal system*

4.6. Comparison of different case studies and their economic analysis

Table 4.9 represents the summarized results obtained from all the cases. The best case with less LCOE, less NPC and 100% renewable penetration is considered for the design of a mini grid. From the simulation results, it is concluded that designing a solar PV, biomass, wind and battery storage system gives better results than all the other configurations of mini-grid design. The LCOE, NPC, renewable penetration and emission are assessed as \$0.195/kWh, \$699,494, 100% and 42.4 kg/year.

Summarized results												
PV (kW)	**Wind (kW)**	**Biomass (kW)**	**DG1 (kW)**	**Battery kWh**	**Converter (kW)**	**NPC (\$)**	**LCOE (\$/kWh)**	**Operating cost (\$/year)**	**Biomass in tonnes**	**Diesel (L)**	**CO_2 (kg/year)**	**Renewable fraction (%)**
60	**20**	**40**	–	**74**	**29.5**	**699,494**	**0.195**	**28,976**	**673**	–	**42.4**	**100**
60	–	50	–	63	37.4	710,434	0.198	35,444	882	–	55.6	100
–	–	50	–	21	6.49	721,780	0.202	42,123	1,118	–	70.5	100
–	40	40	–	131	26.4	808,092	0.226	33,067	701	–	44.2	100
–	–	60	–	–	–	825,536	0.230	48,598	1,224	–	77.2	100
60	–	60	–	–	32.4	862,553	0.241	45,759	1,109	–	69.9	100
–	20	60	–	–	–	948,281	0.265	50,042	1,157	–	72.9	100
60	20	60	–	–	31.5	968,508	0.270	46,155	1,037	–	65.4	100
60	60	–	40	312	30.6	1.56 M	0.435	67,854	–	28,594	75,292	67
–	60	–	50	275	28.7	1.91 M	0.532	95,004	–	47,934	12,227	49.4
60	–	–	50	70	37.5	2.03 M	0.566	120,876	–	71,455	188,164	22.9
–	–	–	50	21	6.49	2.42 M	0.676	152,109	–	91,924	242,066	0
60	60	–	60	–	32	2.43 M	0.679	128,701	–	72,357	190,540	32.2
60	–	–	60	–	34.8	2.48 M	0.691	150,565	–	89,689	236,180	11.2
–	20	–	60	–	–	2.61 M	0.730	158,280	–	93,243	245,540	9.95
–	–	–	60	–	–	2.63 M	0.734	165,517	–	99,482	261,968	0

Table 4.9. *Summarized optimum results*

4.7. Summary

This chapter presents the design and feasibility analysis of a mini-grid for the Denmark location – solar PV, wind, biomass, DG and battery storage power the proposed mini-grid. Different configurations are analyzed, and the results obtained are presented. The best configuration was selected for the design of the mini-grid.

4.8. Conclusion

This study conducted a thorough techno-economic analysis of hybrid renewable energy systems for off-grid community electrification in Aalborg, Denmark, using HOMER Pro software. Among eight evaluated configurations, the solar PV (60 kW) + wind (20 kW) + biomass (40 kW) + battery storage (74 kWh) system emerged as the optimal solution, demonstrating superior performance across all metrics. This winning configuration achieved the lowest LCOE ($0.195/kWh) and NPC ($699,494) while maintaining 100% renewable penetration and exceptional reliability, with a LPSP of just 0.023%. The system's environmental benefits were equally impressive, producing only 42.2 kg of CO_2 emissions annually. These findings have significant implications for renewable energy deployment in remote communities, particularly in cold climates. The study recommends policy measures to strengthen biomass supply chains and initial deployment subsidies to overcome battery storage cost barriers. Future research directions include integrating advanced forecasting techniques and exploring green hydrogen for long-term energy storage. The proposed hybrid system serves as a replicable model for sustainable off-grid electrification, offering a practical pathway to 100% renewable energy communities while addressing real-world challenges of resource variability and energy management. This work contributes valuable insights to the global transition toward decentralized, clean energy systems, demonstrating that technically robust and economically feasible solutions are achievable today.

4.9. References

Boddapati, V. and Daniel, S.A. (2020). Performance analysis and investigations of grid-connected solar power park in Kurnool, South India. *Energy for Sustainable Development*, 55, 161–169.

Boddapati, V. and Daniel, S.A. (2021). Optimal design and technoeconomic analysis of a microgrid for community, load applications. In *Advances in Clean Energy Technologies*, Baredar, P.V., Tangellapalli, S., Solanki, C.S. (eds). Springer, Singapore.

Boddapati, V., Kumar, T.S., Prakash, N., Gunapriya, B. (2021a). Current droop control of parallel inverters in an autonomous microgrid. *Materials Today Proceedings*, 45, 2034–2039.

Boddapati, V., Nandikatti, A.S.R., Daniel, S.A. (2021b). Techno-economic performance assessment and the effect of power evacuation curtailment of a 50 MWp grid-interactive solar power park. *Energy for Sustain. Dev.*, 62(1), 16–28.

Boddapati, V., Sree Ram Nandikatti, A., Arul Daniel, S. (2021c). Design and feasibility analysis of a solar PV array installation during the construction of high-rise residential buildings. *J. Energy Sys.*, 5(2), 60–79.

Boddapati, V., Rakesh Kumar, A., Arul Daniel, S., Padmanaban, S. (2022a). Design and prospective assessment of a hybrid energy-based electric vehicle charging station. *Sustainable Energy Technologies and Assessments*, 53(A), 102389.

Boddapati, V., Kumar, A.R., Prakash, D.B., Daniel, S.A. (2022b). Design and feasibility analysis of a solar PV and biomass-based electric vehicle charging station for metropolitan cities (India). *Distributed Generation & Alternative Energy Journal*, 37(3), 793–818.

Ghasemi, A., Asrari, A., Zarif, M., Abdelwahed, S. (2013). Techno-economic analysis of stand-alone hybrid photovoltaic–diesel–battery systems for rural electrification in eastern part of Iran – A step toward sustainable rural development. *Renewable and Sustainable Energy Reviews*, 28, 456–462. doi: 10.1016/j.rser.2013.08.011.

Jayanth, K.G., Boddapati, V., Geetha, R.S. (2018). Comparative study between three-leg and four-leg current source inverter for solar PV application. In *International Conference on Power, Instrumentation, Control, and Computing (PICC)*. IEEE.

Kumar, R.R., Bharatiraja, C., Udhayakumar, K., Devakirubakaran, S., Sekar, K.S., Mihet-Popa, L. (2023). Advances in batteries, battery modeling, battery management system, battery thermal management, SOC, SOH, and charge/discharge characteristics in EV applications. *IEEE Access*, 11, 105761–105809. doi: 10.1109/ACCESS.2023.3318121.

Kushwah, A., Reina, T.R., Short, M. (2022). Modelling approaches for biomass gasifiers: A comprehensive overview. *Science of the Total Environment*, 834, 155243. doi: 10.1016/j.scitotenv.2022.155243.

Manas, M., Sharma, S., Reddy, K.S., Srivastava, A. (2023). A critical review on techno-economic analysis of hybrid renewable energy resources-based microgrids. *J. Eng. Appl. Sci.*, 70(148). doi: 10.1186/s44147-023-00290-w.

Munjer, M.A., Sheikh, M.R.I., Alim, M.A., Boddapati, V., Musaib, M.A. (2018). Minimization Of THD For multilevel converters with triangular injection approach. In *3rd International Conference for Convergence in Technology (I2CT 2018)*, Pune, 6–8 April. IEEE.

Nethravathi, S. and Murali, V. (2021). A novel residential energy management system based on sequential whale optimization algorithm and fuzzy logic. *Distributed Generation and Alternative Energy Journal*, 37(3), 557–586. doi: 10.13052/dgaej2156-3306.3739.

Shivanaganna, N., Shivamurthy, K.P., Boddapati, V. (2024). Optimal strategy for transition into nearly zero energy residential buildings: A case study. *Energy*, 307, 132742. doi: 10.1016/j.energy.2024.132742.

Soon, J.J. and Low, K.-S. (2015). Optimizing photovoltaic model for different cell technologies using a generalized multidimension diode model. *IEEE Transactions on Industrial Electronics*, 62(10), 6371–6380. doi: 10.1109/TIE.2015.2420617.

5

Impact of Field Degradation Rates on the Levelized Cost of Energy (LCOE) for Solar PV Systems

Computation of levelized cost of energy (LCOE) encompassing realistic field degradation rates is inevitable for accurately determining both the economic feasibility and aging aspects of solar photovoltaic (PV) systems. A miscalculation in LCOE, especially if based on assumed rather than observed degradation rates, can lead to inaccurate cost projections and substantial financial losses for stakeholders and investors. In light of this, this chapter proposes a detailed and data-driven methodology for LCOE computation, explicitly accounting for failure mode-dependent degradation rates observed under real-world operating conditions. The PV system under investigation includes a 125 kWp rooftop grid-connected system that has been operational for 5–6 years and is located in Bengaluru, India. The chapter begins with an experimental procedure to detect failure modes and measure real-time module output. This is used for development of an analytical model to quantify the system's power degradation rate (R_d). The model captures an average annual R_d of 7.89%, reflecting the cumulative effect of environmental exposure, soiling and material aging. This empirically derived degradation rate of 7.7% is then integrated into the LCOE calculation model to yield a more accurate cost of energy generation, estimated at ₹7.11/kWh. This value is benchmarked against widely used tools such as the System Advisor Model (SAM) and the Cost of Renewable Energy Spreadsheet Tool (CREST), which produced comparable LCOE values of ₹9.04/kWh with RMSE of 0.127 and ₹8.84/kWh, with RMSE of 0.0234, respectively. The analysis confirms the critical relationship: higher degradation rates result in higher LCOE. Ultimately, this chapter underscores the importance of moving away from uniform, assumed degradation rates toward empirically validated, field-specific R_d values. By doing so, planners, policymakers and investors can derive more precise long-term cost projections for aging PV systems, leading to informed financial decisions and risk mitigation. The proposed methodology sets a precedent for robust and realistic LCOE modeling that holistically incorporates the dynamic performance of solar assets in the field.

Chapter written by Almas and Sivasankari SUNDARAM.

For a color version of all figures and tables in this chapter, see www.iste.co.uk/sundaram/energysystems.zip.

5.1. Introduction

The 21st century marks a pivotal era in the transformation of global energy systems, driven by urgent environmental concerns, growing energy demands and the imperative to decarbonize. Among various renewable technologies, solar PV energy has emerged as a cornerstone of clean energy strategies due to its scalability, declining costs and relevance across geographies. However, while solar PV adoption has accelerated rapidly, the reliability, economic viability and long-term performance of these systems, particularly in rooftop deployments, remain under continuous scrutiny. To ensure optimal utilization and return on investment, it becomes crucial to move beyond installation metrics and focus on operational performance, degradation trends and realistic cost evaluations over time.

5.1.1. *Global and Indian solar transition*

Concerns over climate change, energy security and sustainability have driven a global energy transition, marked by a significant shift from fossil fuel-based resources to renewable energy sources. Among these, solar PV systems have gained substantial traction due to their modularity, declining cost and suitability for decentralized applications. Globally, countries are making bold commitments to solar energy, aligning with carbon neutrality and sustainable development goals.

India, with its tropical climate and abundant solar resources, stands at the forefront of this transition. The launch of the Jawaharlal Nehru National Solar Mission (JNNSM) in 2010 catalyzed solar adoption across the country. As of 2024, India's total solar installed capacity has crossed 70 GW, with rooftop PV contributing significantly in urban and semi-urban settings. The annual growth rate of grid-connected PV installations in India is estimated at 24% (Anbazhagan et al. 2021), underscoring the country's commitment to renewable energy adoption.

5.1.2. *The rise and challenges of rooftop PV*

Rooftop PV installations have emerged as a key segment, particularly for residential, commercial and industrial consumers aiming to reduce electricity bills and improve energy independence (Gernaat et al. 2020). However, the rooftop sector faces unique challenges compared to utility-scale PV (Asif et al. 2025). These include higher variability in installation quality, non-uniform maintenance practices, environmental soiling (dust, bird droppings), shading issues and lack of real-time performance monitoring (Aïssa et al. 2023). Unlike ground-mounted plants, rooftop systems are often perceived as "install and forget" models. This passive management approach causes lag in performance optimization and predictive maintenance.

5.1.3. *Need for realistic performance assessment*

Solar PV systems have become an install-and-forget model lagging behind decisive monitoring, model quantification and performance optimization. The true potential of PV modules, which form the most integral part of any solar PV system, depends on their on-field degradation and failure modes. This is influenced by their exposure to various environmental conditions, affecting power reliability and economic viability. To protect the cash flows made by the system owner and support the role of the grid system operator, accurate knowledge on the factors affecting PV system performance – primarily PV module operation and its impact on the LCOE economic indicator – is crucial (Rana and Jindal 2025). Power loss and financial hazards are introduced when PV modules deteriorate over time (Poddar et al. 2024).

5.1.4. *PV degradation: mechanisms and impact*

Consequently, it is imperative to comprehend the mechanisms underlying module degradation and their impact on LCOE (Limmanee et al. 2017). Computation of LCOE involving realistic system metrics – especially failure-dependent operational degradation rate (R_d) – serves as a true computing tool under real-time on-field conditions. Incorporating degradation rates into LCOE calculation offers a more realistic framework for evaluating the long-term costs of aging PV systems. Module degradation is influenced by factors such as ultraviolet (UV) exposure, thermal cycling, humidity ingress and potential-induced degradation (PID) (Yang et al. 2025). These effects vary based on module technology, installation practices and environmental stressors, and may not always be linear.

5.1.5. *Role and limitations of LCOE*

The reported methodologies for LCOE calculation typically take into account the total energy and related expenses over a project's lifetime, assuming a constant rate of power degradation irrespective of changing field conditions. This presents a problem for emerging PV technologies since energy yield deteriorates more quickly than that of established silicon PV modules, which are typically guaranteed to lose not more than 0.7% of their yield annually over the course of their average 25-year lifespan (Nieto-Díaz et al. 2021). As a metric, LCOE encapsulates all costs associated with the development, operation and decommissioning of energy systems, normalized by their total energy output over their operational lifespan

(Kibaara et al. 2020). Its utility in evaluating the feasibility of renewable energy projects has been widely acknowledged, and numerous studies have explored its applications and implications.

Jordan et al. (2016) inferred that degradation rates based on nameplate ratings could be greatly affected. Even after controlling for environment, research that utilized solar simulators and nameplate ratings as a reference showed less variation than comparable studies that employed outdoor measurements. This may be related to the confounding effects of outside soiling and measurement uncertainty. The LCOE is significantly impacted by modeling these nonlinearities, whether they arise at the start or finish of the PV lifecycle.

5.1.6. *Gap in literature and industry practice*

Despite the widespread use of LCOE as a decision-making tool, the lack of integration of field-derived degradation data into its computation leaves room for misestimations in cost–benefit analysis. Most LCOE models are generic, assuming ideal or lab-like conditions, which fail to capture real-world variability. As a result, investments based on such models may lead to unexpected operational and maintenance costs, or misjudgment in project feasibility. Thus, there is a critical need for methodologies that integrate real-time, on-site degradation metrics into economic performance evaluations.

5.1.7. *Scope and structure of the chapter*

This chapter focuses on bridging these gaps by experimentally quantifying the degradation rate and integrating the on-field R_d into the widely employed LCOE model. By leveraging on-field performance characteristics, the chapter provides a comprehensive approach to assessing the economic performance of rooftop solar PV systems. The objectives of the present investigation are as follows:

– to monitor and quantify the real-time degradation rate (R_d) of rooftop PV modules under Indian climatic conditions;

– to develop a field-integrated LCOE model that reflects true economic performance over time;

– to assess how variations in R_d influence energy yield and financial metrics such as payback period and internal rate of return (IRR);

– to compare the enhanced LCOE framework with traditional models and to illustrate the impact of incorporating degradation data.

The chapter is structured as follows: section 5.2 will delve into the literature on PV degradation and LCOE computation techniques. Section 5.3 will describe the experimental setup, monitoring parameters and data collection procedures. Section 5.4 will elaborate on the modeling framework, including the LCOE formulation and R_d integration. Section 5.5 will present results, sensitivity analysis and comparative evaluation. Section 5.6 will conclude with insights for researchers, policymakers and solar stakeholders.

5.2. Literature on PV degradation and LCOE computation techniques

The performance degradation of PV modules significantly affects both energy yield and financial returns over their operational lifetime (Carigiet et al. 2021). Understanding the mechanisms, estimation techniques and economic modeling approaches is essential for improving decision-making and financial planning in solar energy investments (Ameur et al. 2022).

5.2.1. *PV degradation: mechanisms and field observations*

PV module degradation is caused by a combination of material aging and environmental stressors (Omazic et al. 2019). Mechanisms such as light-induced degradation (LID), moisture ingress, PID and thermomechanical stress result in physical defects such as delamination, encapsulant discoloration and microcracks. These, in turn, manifest as hot spots, voltage drops and long-term energy loss.

Studies such as Kaldellis et al. (2014) observed that degradation rates can exceed 1% per annum in coastal or arid zones due to salinity and thermal cycling. Ricardo et al. (2022) emphasized the role of urban pollution and soiling in accelerating performance loss in rooftop PV installations, particularly in high-density cities.

Furthermore, empirical data from long-term monitoring such as those from the National Renewable Energy Laboratory (NREL) PV Fleet Performance Database confirm that nonlinear and environment-specific degradation patterns are more prevalent than the manufacturer-stated uniform linear decline. This divergence has prompted the need for more location-specific degradation modeling.

5.2.2. *Approaches to degradation estimation*

Broadly, degradation rate (R_d) estimation methods fall into three categories:

– *Empirical on-field monitoring*: techniques such as I–V curve tracing (IEC 60891), string-level performance tracking and electroluminescence imaging are used to detect early failure signatures and power loss trends in operational systems (Meena et al. 2022; Augusto et al. 2022; Schuck De Oliveira et al. 2024).

– *Accelerated lifetime testing*: standards such as IEC 61215 involve high-temperature and high-humidity stress testing. While useful for product validation, they often do not correlate directly with field degradation profiles (Boulhidja et al. 2020; Kim et al. 2021).

– *Analytical and AI-driven models*: recent work by Dhimish et al. (2017) applied convolutional neural networks to detect and quantify power loss from EL images. Similarly, Chatterjee et al. (2024) proposed a hybrid machine learning model integrating meteorological variables and visual defect classification for R_d prediction.

5.2.3. *LCOE modeling and its limitations*

The LCOE remains the dominant metric for evaluating the cost-competitiveness of PV systems. However, its accuracy heavily depends on assumptions regarding system lifespan, degradation, discount rates and maintenance costs. Traditional models, such as those outlined by the IEA-PVPS Task 13 framework, assume a constant annual degradation rate (Weiß et al. 2021).

This assumption often fails to capture early-year performance dips due to installation issues or long-term impacts of field aging. Studies like Nieto-Díaz et al. (2021) have shown that underestimating R_d by even 0.5% per year can misrepresent project viability by more than 10% in net present value (NPV).

Furthermore, emerging rooftop PV installations that are often user-managed and exposed to highly variable conditions require dynamic LCOE models that incorporate failure-specific R_d and evolving maintenance profiles (Poornima et al. 2025). Tools like NREL’s System Advisor Model (SAM) are starting to offer modularity for such inputs but still rely on user assumptions for R_d (Barva et al. 2024).

5.2.4. *Research gap and motivation*

While significant progress has been made in diagnosing PV degradation and refining LCOE computations, an integrated framework that combines long-term field data, failure-mode insights and machine learning-based R_d estimation remains largely unexplored. Most existing studies either focus on lab conditions or offer generic, non-specific degradation modeling.

This research aims to address these gaps by:

– deriving a failure-mode-specific R_d from on-field rooftop experimentally analyzed data;

– using nonlinear degradation trends derived from environmental and electrical parameters;

– integrating the observed R_d into an LCOE model to assess economic viability over time.

5.2.5. *Objectives of the study*

The objectives of this research are formulated based on the gaps identified in conventional LCOE modeling practices and the need to realistically capture the degradation behavior of rooftop solar PV systems. The following aims guide the experimental and analytical framework of this study:

1) To integrate the realistic, failure-mode-dependent R_d into the LCOE model to compute levelized cost values that reflect true performance loss over time. This objective modifies the conventional LCOE formulation by replacing the constant annual degradation rate with the observed, time-series-based R_d.

2) To evaluate the long-term financial impact of using actual degradation values versus assumed values on LCOE, NPV and IRR. This analysis helps determine how economic assessments change when the degradation profile is tailored to local environmental conditions and failure modes.

Each objective is designed to contribute both practically and scientifically to the emerging field of PV system performance modeling, offering a cost-effective yet accurate alternative to expensive and time-intensive accelerated field testing. The following section will describe the experimental setup, data acquisition methods, sensor suite and processing techniques employed to fulfil the first objective.

5.3. Experimental setup and data acquisition framework

To achieve a realistic and failure-mode-specific estimation of degradation rate factor (R_d), an extensive on-field experimental investigation was conducted on a rooftop solar PV system, as shown in Figure 5.1.

This section outlines the system characteristics, instrumentation used and the procedure adopted for data acquisition and preliminary preprocessing.

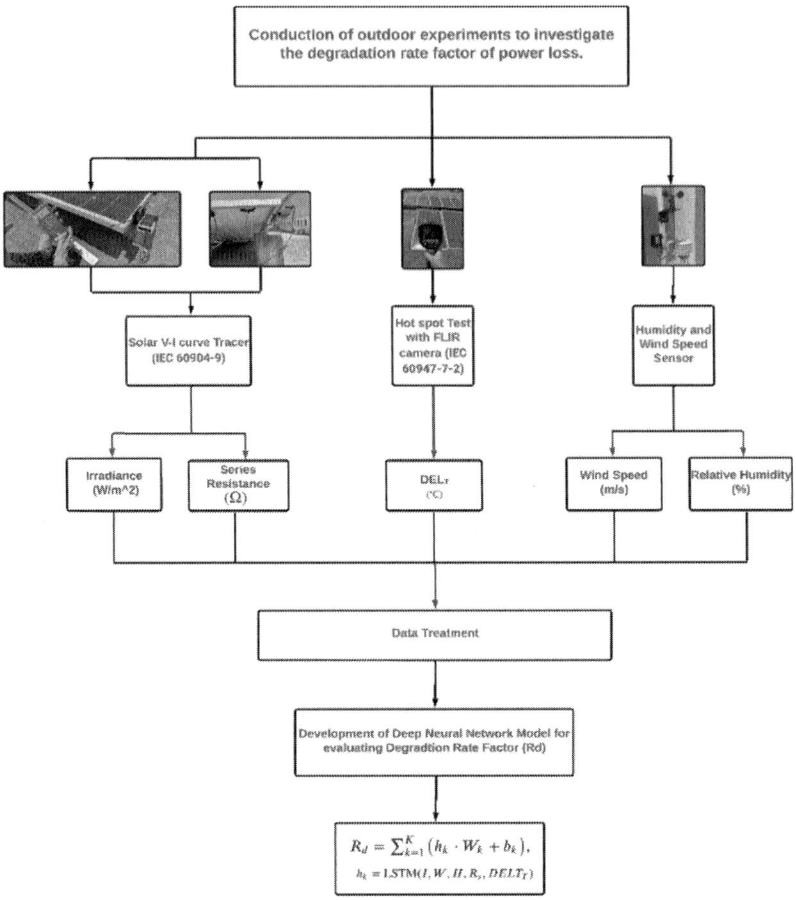

Figure 5.1. *Methodology for evaluating on-field degradation rate factor*

5.3.1. *Description of the PV system*

The experimental analysis was conducted on a 125 kWp rooftop grid-connected PV plant located at 13.1596° N and 77.6362° E in Bengaluru, Karnataka, India. The plant, commissioned more than five years ago, consists of polycrystalline silicon modules rated at 325 Wp each, manufactured by Vikram Solar. The system is connected to a three-phase inverter with a rated capacity of 50 kW.

The modules are installed facing south at a fixed tilt angle of 13°. The geographical location provides approximately 2,365 sunshine hours annually, with moderate variation in seasonal temperatures, making it suitable for long-term performance degradation studies.

5.3.2. *Module selection and visual diagnostics*

A total of 13 modules from a string of 20 were shortlisted for detailed assessment based on visual inspection. The selection was done as per IEC 61215-2 protocols, focusing on identifying signs of degradation and failure modes such as:

– discoloration and delamination;

– snail trails and cell cracks;

– backsheet blistering;

– junction box oxidation;

– frame defects and glass breakage.

The aim was to isolate modules that had clear, visually observable defects likely to contribute to nonlinear degradation.

5.3.3. *Instrumentation and measurement protocol*

The following instruments were deployed for field data collection:

– *Solar V–I curve tracer (Seaward PV200)*: this is used to measure I–V characteristics of each module at solar noon. It also captures key electrical parameters such as open-circuit voltage (V_{oc}), short-circuit current (I_{sc}), maximum power point voltage and current (V_{mpp} and I_{mpp}), and derived maximum power (P_{max}).

– *Pt100 RTD sensor and surface temperature probe*: this is used for monitoring module backsheet and ambient temperatures.

– *Irradiance sensor*: this is mounted on module plane to record real-time solar irradiance (W/m^2).

– *Anemometer and hygrometer*: this is used for capturing wind speed (m/s) and relative humidity (%).

– *Infrared thermal camera (FLIR E5-XT)*: this is employed to detect hot spots and estimate the temperature differential (DELT) across the module surface.

Measurements were taken around solar noon under clear sky conditions to minimize variability and align closely with standard test conditions (STC).

5.3.4. *Data acquisition duration and parameters logged*

The data collection campaign spanned twelve months – from February 2022 to January 2023 – covering seasonal variations in ambient conditions and irradiance. The parameters logged per observation included:

– *electrical*: V_{oc}, I_{sc}, V_{mpp}, I_{mpp} and P_{max};

– *environmental*: irradiance, ambient temperature, wind speed and relative humidity;

– *thermal diagnostics*: DELT (temperature differential across module surface);

– *derived*: series resistance (R_s), corrected power at STC (PCR).

These parameters were later used to compute the power degradation factor (R_d) and validate failure mode impact.

5.3.5. *Data preprocessing*

The raw data underwent the following preprocessing steps before being fed into the Rd estimation model:

– *Outlier removal*: data points with irradiance below 200 W/m^2 were discarded to avoid skew from low insolation conditions.

– *Outlier correction*: abnormal values were replaced using weighted averages of nearby significant observations.

– *Normalization*: all input parameters were normalized using Min–Max scaling to facilitate machine learning model convergence and reduce prediction error.

This comprehensive field-based experimental protocol ensures that the subsequent degradation modeling captures true operational behavior and failure-induced losses, rather than relying on generic manufacturer data or synthetic test conditions. Furthermore, a long short-term memory (LSTM) model was developed using the same dataset to predict degradation rate factor (R_d). The consistency between the model's outputs and empirically observed R_d values further validates the reliability of the field-derived degradation behavior.

To accurately capture the degradation rate factor (Rd) for rooftop solar PV systems, a nonlinear, machine learning-based model was developed using LSTM architecture. Model 1 employs five key input parameters: irradiance (I), wind speed (W), relative humidity (H), series resistance (Rs) and the temperature differential (DELT) between maximum and minimum module temperatures. These inputs were selected based on their statistical significance (p-values < 0.05) and observed field influence on degradation mechanisms.

5.3.6. *Development of R_d model*

The model was trained using one year of field-acquired data (February 2022–January 2023) from a 125 kWp rooftop grid-connected PV system located at 13.1596° N, 77.6362° E in Bengaluru, India. Data preprocessing involved outlier removal (irradiance < 200 W/m^2 excluded), Min–Max normalization and validation against visual inspection and I–V tracer measurements.

This achieved a root mean square error (RMSE) of 0.0019 during testing and 0.000612 during training, indicating strong predictive accuracy with a closeness of prediction to actual values of 9.0%. Incorporation of the DEL_T parameter improved accuracy by approximately 5.7%. The empirically determined annual degradation rate was 7.7%/year, as illustrated in Figure 5.2, which was subsequently integrated into the LCOE framework for more realistic economic assessment.

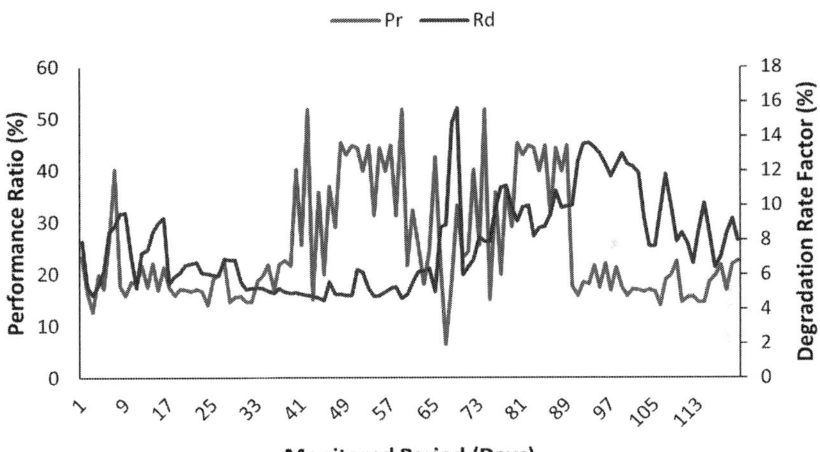

Figure 5.2. *Variation of performance ratio (%) and degradation rate factor (R_d) for the plant under investigation*

5.4. LCOE formulation and integration of field-derived R_d

To evaluate the true economic impact of degradation in PV systems, an LCOE framework was applied using the experimentally obtained degradation rate factor (R_d). This section outlines the calculation methodology, key financial assumptions and comparative results of LCOE, NPV as seen in equation [5.1] and IRR using both field-derived and assumed degradation rates.

$$NPV = -II + \sum_{t}^{PL} \frac{EG_t - OE_t}{(1+DR)^2} \quad [5.1]$$

5.4.1. *LCOE formulation*

The LCOE is calculated as the ratio of the present value of total lifecycle costs to the present value of energy generation, expressed as in equations [5.2] and [5.3].

$$LCOE = \frac{II + \sum_{t=1}^{PL} OE_t}{\sum_{t=1}^{PL} \frac{EG_t}{(1+DR)^t}} \quad [5.2]$$

$$EG_t = EG_0 \times (1 - R_d)^t \quad [5.3]$$

where – II is initial investment cost, OE_t is operating expenses in year, EG_t is energy generated in year, DR is discount rate, PL is project lifetime (25 years considered), R_d is degradation rate and EG_0 is base year energy generation.

The degradation-adjusted annual energy generation is estimated using equation [5.3], where Rd is the degradation rate (field-observed: 7.7%, industry-assumed: 5%). When NPV is calculated, the following formulation is used:

When NPV=0, IRR is the discount rate (denoted as DR) that makes the present value of future cash flows equal to the initial investment as seen in equation [5.4].

$$0 = -II + \sum_{t}^{PL} \frac{EG_t - OE_t}{(1+IRR)^2} \quad [5.4]$$

5.4.2. *Financial assumptions*

Financial assumptions are as follows:

– Initial investment (II): ₹5,810,000.

– Annual O&M cost: ₹21,349.55.

– Project lifetime: 25 years.

– Discount rate: 8%.

– Base year energy generation: 147,086 kWh.

5.4.3. *Comparative evaluation*

Using equations [5.1]–[5.3], and validated through both SAM and CREST, the economic indicators were derived under different R_d scenarios, as seen in Table 5.1.

Sr. No.	Economic indicators	Estimation with field Rd 7.7% (using equations [5.1]–[5.3])	Estimation with assumed Rd 5% (using equations [5.1]–[5.3])	Estimation with field Rd 7.7% with System Advisory Model (SAM)	Estimation with field Rd 7.7% with Cost of Renewable Energy Spreadsheet Tool (CREST)
1	LCOE – Discounted Energy Generation (kWh) (DR) – Discounted Operating Expenses (OE_t)	**₹7.11/kWh** – 36,735.93 kWh/year – ₹21,349.55/year	**₹6.02/kWh** – 43,424.14 kWh/year – ₹21,349.55/year	₹9.04/kWh	₹8.84/kWh
2	NPV – Initial Investment (II) – Annual Energy Generation (EG_t)	**₹41,982.05** – ₹5,810,000.00 – 147,086 kWh/year	**₹65,164.11** – ₹5,810,000.00 – 147,086 kWh/year	**₹42,528.3**	**₹6,676,042**
3	IRR	9.34%	12.09%	12.5%	11.34%

Table 5.1. *Comparison of economic indicators*

5.4.4. *Insights from field-based R_d integration*

The integration of field-derived R_d significantly alters the economic landscape of the PV project. While a lower assumed R_d of 5% yields more optimistic metrics (e.g. IRR of 12.09% and LCOE of ₹6.02/kWh), the realistic R_d of 7.7% increases LCOE to ₹7.11/kWh and reduces IRR to 9.34%, highlighting potential overestimations in traditional financial planning.

Simulations using SAM and CREST tools further validate this discrepancy, producing LCOE values closer to ₹9/kWh under real-world degradation, especially when module failure modes are not adequately addressed.

These findings affirm the critical importance of accurately estimating R_d for realistic economic projections. The use of field-derived degradation rates not only improves accuracy in lifecycle costing but also reinforces the broader reliability of performance forecasting models

5.5. Results and discussion

The evaluation of economic indicators under different degradation scenarios provides key insights into the financial viability of the 125 kWp grid-connected PV system. By incorporating the field-observed degradation rate (7.7%) into the economic models, a more realistic depiction of long-term performance and cost was achieved.

5.5.1. *Impact of degradation rate on economic indicators*

As observed in Table 5.1 (section 4.3), the LCOE for the PV system increases from ₹6.02/kWh (with assumed R_d of 5%) to ₹7.11/kWh when the true degradation of 7.7% is used. This shift underscores how over-optimistic assumptions in R_d can substantially underestimate the cost of energy. Similarly, the IRR drops from 12.09% to 9.34%, and NPV reduces from ₹65,164.11 to ₹41,982.05 under the field degradation scenario. The decrease in discounted energy generation from 43,424.14 kWh/year (assumed R_d) to 36,735.93 kWh/year (field R_d) further quantifies the impact of degradation on lifecycle energy output and revenue potential. This reflects the compounding effect of even marginal increases in R_d over the system's operational lifetime.

5.5.2. *Benchmarking with simulation tools*

To verify the validity of the in-house calculations, the same field R_d (7.7%) and system parameters were applied in SAM and CREST. SAM yielded an LCOE of ₹9.04/kWh while CREST gave ₹8.84/kWh, both in close agreement with the analytical value of ₹7.11/kWh, especially when accounting for methodological differences and tool-specific assumptions.

These results reinforce the robustness of the field-R_d–based economic analysis and demonstrate that third-party tools, when fed accurate degradation data, can replicate real-world performance estimations effectively.

5.5.3. *Implications for PV financial planning*

This analysis illustrates the sensitivity of economic viability to accurate degradation profiling. An underestimation of R_d can lead to inflated IRR and NPV projections, risking overcommitment of capital and misaligned expectations. As such, integrating site-specific or field-measured degradation data into financial assessments should be considered a best practice for utility-scale and distributed PV systems alike.

5.6. Conclusion and future scope

This chapter underscores the critical importance of integrating experimentally derived degradation rate factors (R_d) into PV economic analysis. By leveraging field-monitored data, including I-V characteristics, thermographic observations and environmental conditions, a realistic R_d of 7.7% was established for a 125 kWp rooftop PV installation. The incorporation of this R_d into LCOE computations revealed a significant deviation from conventional financial estimates based on ideal or assumed degradation rates (typically 5%).

Key economic indicators LCOE, NPV and IRR were recalculated using both analytical methods and benchmarked tools such as SAM and CREST.

The results confirmed that assuming a lower degradation rate can lead to underestimation of LCOE and overestimation of financial viability. Specifically, the LCOE increased from ₹6.02/kWh to ₹7.11/kWh when shifting from assumed to field-derived R_d, while IRR dropped from 12.09% to 9.34%. SAM and CREST tools corroborated this trend, validating the relevance of field-integrated economic modeling.

The chapter highlights that accurate field data are not just beneficial for performance forecasting but is vital for robust financial planning. It discourages reliance on assumed degradation models in economic assessments and advocates for empirical integration to avoid project-level risks.

5.7. Future scope

– *Long-term monitoring*: extending the observation period beyond one year could further refine R_d estimation, capturing seasonal and long-term effects.

– *Module-specific degradation mapping*: disaggregating R_d at the module or string level may improve fault detection and predictive maintenance strategies.

– *Integration with ML models*: the empirical R_d dataset could serve as training input for LSTM or other deep learning models to forecast long-term degradation trends and optimize maintenance schedules.

– *Policy and investment implications*: incorporating field-based R_d into financial feasibility reports can aid policymakers and investors in better evaluating the risk-reward profile of PV assets.

5.8. References

Aïssa, B., Isaifan, R.J., Figgis, B.W., Abdallah, A.A., Bachour, D., Perez-Astudillo, D., Sanfilippo, A., Lopez-Garcia, J., Bermudez Benito, V. (2023). A comprehensive review of a decade of field PV soiling assessment in QEERI's outdoor test facility in Qatar: Learned lessons and recommendations. *Energies*, 16(13), 5224. doi: 10.3390/en16135224.

Ameur, A., Berrada, A., Bouaichi, A., Loudiyi, K. (2022). Long-term performance and degradation analysis of different PV modules under temperate climate. *Renewable Energy*, 188, 37–51. doi: 10.1016/j.renene.2022.02.025.

Anbazhagan, G., Navamani, D., Anbazhagan, L., Muthusamy, S., Pandiyan, S., Panchal, H., Ramachandran M, Sundarajan S.C.M., Sadasivuni, K.K. (2021). Performance investigation of 140 kW grid connected solar PV system installed in southern region of India – A detailed case study and analysis. *Energy Sources, Part A: Recovery, Utilization, and Environmental Effects*, 45(4), 10472–10486. doi: 10.1080/15567036.2021.2008064.

Asif, M., Sharieff, R., Olawale, M., Imran Khan, M. (2025). Unlocking the potential of unregulated rooftops for solar PV on residential buildings: Identifying and addressing key challenges. *Energy Nexus*, 18, 100447. doi: 10.1016/j.nexus.2025.100447.

Augusto, A., Killam, A., Bowden, S.G., Wilterdink, H. (2022). Measuring outdoor I–V characteristics of PV modules and systems. *Progress in Energy*, 4(4), 042006. doi: 10.1088/2516-1083/ac851c.

Barva, K., Singhal, J., Joshi, S. (2024). Analyzing 5kW rooftop residential photovoltaic system using system advisor model. In *2024 IEEE 9th International Conference for Convergence in Technology (I2CT)*, Pune. doi: 10.1109/i2ct61223.2024.10544321.

Boulhidja, S., Mellit, A., Voswinckel, S., Lughi, V., Ciocia, A., Spertino, F., Massi Pavan, A. (2020). Experimental evidence of PID effect on CIGS photovoltaic modules. *Energies*, 13(3), 537. doi: 10.3390/en13030537.

Carigiet, F., Brabec, C.J., Baumgartner, F.P. (2021). Long-term power degradation analysis of crystalline silicon PV modules using indoor and outdoor measurement techniques. *Renewable and Sustainable Energy Reviews*, 144, 111005. doi: 10.1016/j.rser.2021.111005.

Chatterjee, S., Khan, P.W., Byun, Y.-C. (2024). Recent advances and applications of machine learning in the variable renewable energy sector. *Energy Reports*, 12, 5044–5065. doi: 10.1016/j.egyr.2024.09.073.

Conceição, R., González-Aguilar, J., Merrouni, A.A., Romero, M. (2022). Soiling effect in solar energy conversion systems: A review. *Renewable and Sustainable Energy Reviews*, 162, 112434, doi: 10.1016/j.rser.2022.112434.

Dhimish, M., Holmes, V., Dales, M., Mather, P., Sibley, M., Chong, B., Zhang, L. (2017). The impact of cracks on the performance of photovoltaic modules. In *2017 IEEE Manchester PowerTech*, Manchester, 1–6. doi: 10.1109/PTC.2017.7980824.

Gernaat, D.E.H.J., de Boer, H.-S., Dammeier, L.C., van Vuuren, D.P. (2020). The role of residential rooftop photovoltaic in long-term energy and climate scenarios. *Energy*, 206, 118024. doi: 10.1016/j.energy.2020.118024.

Jordan, D.C. and Kurtz, S.R. (2013). Photovoltaic degradation rates: An analytical review. *Progress in Photovoltaics: Research and Applications*, 21(1), 12–29. doi: 10.1002/pip.1182.

Jordan, D.C., Kurtz, S.R., VanSant, K., and Newmiller, J. (2016). Compendium of photovoltaic degradation rates. *Progress in Photovoltaics: Research and Applications*, 24(7), 978–989. doi: 10.1002/pip.2744.

Kaldellis, J.K. Kapsali, M., Kavadias, K.A. (2014). Temperature and wind speed impact on the efficiency of PV installations. Experience obtained from outdoor measurements in Greece. *Renewable Energy*, 66, 612–624. doi: 10.1016/j.renene.2013.12.041

Kibaara, S., Murage, D.K., Musau, P., Saulo, M.J. (2020). Analysis of the levelized cost of electricity (LCOE) of solar PV systems considering their environmental impacts on biodiversity. In *2020 6th IEEE International Energy Conference (ENERGYCon)*, Gammarth. doi: 10.1109/ENERGYCon48941.2020.9236590.

Kim, J., Rabelo, M., Padi, S.P., Yousuf, H., Cho, E.-C., Yi, J. (2021). A review of the degradation of photovoltaic modules for life expectancy. *Energies*, 14(14), 4278. doi: 10.3390/en14144278.

Limmanee, A., Songtrai, S., Udomdachanut, N., Songpakit, K., Sato, Y., Nakaishi, M., Kittisontirak, S., Sriprapha, K., Sakamoto, Y. (2017). Degradation analysis of photovoltaic modules under tropical climatic conditions and its impacts on LCOE. *Renewable Energy*, 102, 199–204. doi: 10.1016/j.renene.2016.10.052.

Meena, R., Kumar, M., Gupta, R. (2022). Investigation of dominant degradation mode in field-aged photovoltaic modules using novel differential current-voltage analysis approach. *Progress in Photovoltaics*, 30(11), 1312–1324. doi: 10.1002/pip.3580.

Mekhilef, S., Saidur, R., Safari, A., Muhida, R. (2010). A review on solar energy use in industries. *Energy Conversion and Management*, 52(1), 308–317. doi: 10.1016/j.enconman.2010.07.011.

Nieto-Díaz, B.A., Crossland, A.F., Groves, C. (2021). A levelized cost of energy approach to select and optimise emerging PV technologies: The relative impact of degradation, cost and initial efficiency. *Applied Energy*, 299, 117302. doi: 10.1016/j.apenergy.2021.117302.

Omazic, A., Oreski, G., Halwachs, M., Eder, G.C., Hirschl, C., Neumaier, L., Pinter, G., Erceg, M. (2019). Relation between degradation of polymeric components in crystalline silicon PV module and climatic conditions: A literature review. *Solar Energy Materials and Solar Cells*, 192, 123–133. doi: 10.1016/j.solmat.2018.12.027.

Poddar, S., Rougieux, F., Evans, J.P., Kay, M., Prasad, A.A., Bremner, S.P. (2024). Accelerated degradation of photovoltaic modules under a future warmer climate. *Progress in Photovoltaics: Research and Applications*, 32(7), 456–467. doi: 10.1002/pip.3788.

Poornima, P.U., Dhineshkumar, K., Chunduri, K.K, Sumana, S., Rama Sundari, M.V., Sivaraman, P., Mohammed, S., Rajaram, A. (2025). Optimising rooftop photovoltaic adoption in urban landscapes: A system dynamics approach for sustainable energy transitions. *Biomedical Signal Processing and Control*, 100, 107071. doi: 10.1016/j.bspc.2024.107071.

Rana, G.S. and Jindal, R. (2025). Factors affecting solar levelized cost of electricity in India and policy recommendations. *Energy and Climate Change*, 6, 100207. doi: 10.1016/j.egycc.2025.100207.

Ricardo, C, José, G.-A., Ahmed, A., M., Manuel, R. (2022). Soiling effect in solar energy conversion systems: A review. *Renewable and Sustainable Energy Reviews*, 162, 112434, doi: 10.1016/j.rser.2022.112434.

Sanz-Saiz, C., Polo, J., Martín-Chivelet, N., Alonso-García, M.D.C. (2022). Soiling loss characterization for photovoltaics in buildings: A systematic analysis for the Madrid region. *Journal of Cleaner Production*, 332, 130041, doi: 10.1016/j.jclepro.2021.130041.

Schuck De Oliveira, F., Perin Gasparin, F., Detzel Kipper, F., Krenzinger, A. (2024). An alternative method to IEC 60891 standard for I–V curve correction based on a single measurement. *Solar Energy*, 284, 113067, doi: 10.1016/j.solener.2024.113067.

Weiß, K.-A., Kaaya, I., Oreski, G., Bruckman, L.S., French, R.H., Tanahashi, T. (2021). Needs, challenges and approaches for new service life estimation models for PV modules – Results from IEA PVPS task 13 subtask 1.4. In *Proceedings of the ISES Solar World Congress 2021*. International Solar Energy Society. doi: 10.18086/swc.2021.03.01.

Yang, H., Yin, Y., Abu-Siada, A. (2025). A comprehensive review of solar panel performance degradation and adaptive mitigation strategies. *IET Control Theory & Applications*, 19(1). doi: 10.1049/cth2.70040.

6

Harnessing the Future: The Critical Role of Energy System Models in Shaping Sustainable and Resilient Global Energy Landscapes

Future global energy systems are contingent on innovation and the application of cutting-edge energy system models, whose design is quite significant for bringing energy system sustainability and resilience into the production, transmission and utilization of energy. Energy system models allow for diversified energy scenarios to be simulated, inclining toward the combined use of renewable energy, raising energy use efficiency and guaranteeing the security of supply with increasing environmental and geopolitical stress. In this chapter, the central contribution of energy system models in shaping a future for global energy, which is sustainable and resilient is illustrated, with an emphasis on the contribution toward policy formation, technology breakthrough and long-term planning. Statistical evidence indicates that the proportion of the world's renewable energy to total energy use has risen by about 10% for the last 10 years, and it is reported to offer more than 50% of the world's electricity by 2030. Additionally, models also calculate that the combined energy systems have the ability to lower the level of greenhouse gas emissions to 70% by 2050 to meet global climate targets.

This chapter highlights the significance of integrated modeling models, including bottom-up, top-down and hybrid models, to enable the prediction of the effect of policy choices, technological innovations and economic patterns on energy systems. As countries are establishing net-zero goals and striving for more energy resilience, energy system models will be critical in providing a secure, equitable and sustainable energy future.

Chapter written by Venkatesh R and Subramani K.

6.1. Introduction

Energy challenge	Key factors	Impact	Solutions
Global energy demand	Population growth Economic development Increased industrialization	Global energy demand expected to increase by 28% by 2040 (Newell et al. 2016)	Diversification of energy sources Increase energy efficiency
Greenhouse gas emissions	Fossil fuel consumption Electricity generation Transportation	75% of global CO_2 emissions from fossil fuels (Craig and Latham 2024)	Shift to renewable energy sources Carbon capture and storage (CCS)
Renewable energy integration	Solar, wind, hydropower Intermittency of sources Storage issues	Solar and wind accounted for 10% of global energy in 2022 (Ashraf and Sagheer 2025)	Improved storage technologies Grid modernization Incentives for renewable energy
Energy security	Fossil fuel dependence Geopolitical instability Price volatility	60% of global oil production controlled by OPEC (Montant 2025)	Diversify energy sources Develop domestic renewable energy capacity
Energy efficiency	Energy loss in systems Inefficient infrastructure Consumer behavior	Global energy efficiency improvement rate at 1.8% annually (Newell and Iler 2013)	Invest in energy-efficient technologies Improve industrial and residential efficiency
Energy access	Global disparity in energy access Energy poverty in developing countries	759 million people still lack electricity access (Dabla and Goldthau 2023)	Expand energy infrastructure Support off-grid solutions Promote sustainable development goals (SDGs)

Table 6.1. *Key challenges and solutions in the global energy transition*

The global community is at the juncture of addressing its acute problem in how to achieve the increased demand for energy while preserving the environment. As the global population rises and economies grow, there is a growing demand for energy that places resources and infrastructure under pressure. It coincides with increased burning of fossil fuels, which is largely responsible for climatic change and is, arguably, the most immediate world problem (Pfenninger et al. 2014). Discovering the means to satisfy energy requirements without increasing environmental harm is crucial to a sustainable future. There are a number of interrelated issues driving the global energy challenge – as mentioned in Figure 6.1 and with key factors shown in Table 6.1. One of the most critical is lowering greenhouse gas emissions, which primarily result from the generation of electricity, transportation and industry from fossil fuels. To slow down climate change, we must shift to cleaner energy sources such as solar, wind and hydropower. The trouble with these renewable energy sources is that they are unpredictable, have storage problems and require new infrastructure.

Another major concern is energy security. Most nations are dependent on fossil fuels, particularly oil and gas, which can pose threats due to supply shortages and price fluctuations. To have a sound energy supply, we must diversify our sources of energy and enhance infrastructure. To overcome these issues, energy systems need to develop to serve increasing demand while enhancing sustainability. Energy system models simulate different energy scenarios so that policymakers can understand the potential impacts, plan for the future and make good-informed decisions (Keirstead et al. 2012). Through the models, we can better incorporate renewable energy, raise efficiency, promote security and offer energy access to all.

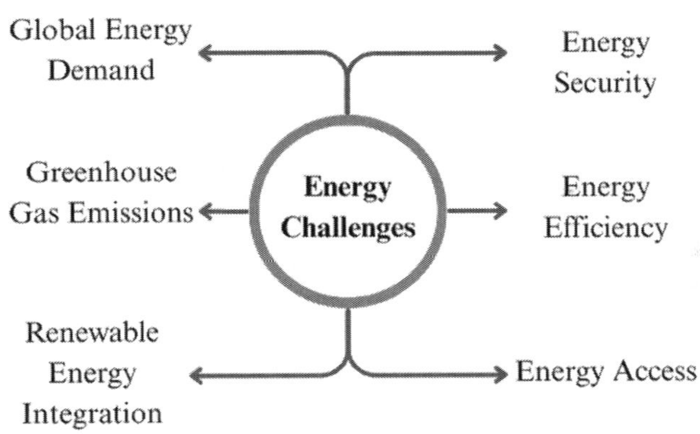

Figure 6.1. *Factors contributing to the global energy challenges*

In this regard, energy system models provide valuable insights and solutions for transitioning to a cleaner, more sustainable, and secure energy future. Table 6.1 outlines the key challenges, their quantitative impact and possible solutions. It helps frame the world energy issue and how energy system models can help address such issues by providing information and guiding policy.

6.1.1. *Overview of global energy challenges*

Challenge	Description	Key issues	Required solutions
Climate change	The burning of fossil fuels for energy production is a major contributor to global greenhouse gas emissions and climate change.	Significant emissions from the energy sector Need for low-carbon alternatives	Transition to renewable energy sources Increased energy efficiency Policy measures to reduce emissions
Rising energy demand	Growing populations, urbanization and industrialization, especially in developing regions, lead to an increasing demand for energy.	Pressure on existing infrastructure Growing need for energy in developing regions	Expansion and modernization of energy infrastructure Efficient energy production and consumption
Energy security	Dependence on fossil fuels, especially oil and gas, exposes countries to geopolitical risks and supply disruptions.	Vulnerabilities in global supply chains Price fluctuations Energy dependencies	Diversification of energy sources Strengthening energy security through renewable energy Improved international cooperation
Renewable energy transition	The shift to solar, wind and hydropower energy presents challenges like energy variability, storage issues and grid integration.	Intermittency of renewable energy Storage limitations Grid modernization needed	Development of advanced energy storage technologies Improved grid management Investment in smart grid technologies
Technological advancements	New technologies like energy storage solutions and grid management systems are needed to optimize energy systems.	High development costs Technological barriers Need for large-scale implementation	Investment in research and development of energy storage Adoption of smart grid and AI-driven grid management systems Policy incentives for innovation

Table 6.2. *Global energy challenges: key issues and solutions*

The global energy environment is faced with a set of interconnected challenges that require immediate attention and innovative solutions. The most pressing among these is climate change, which is primarily caused by the burning of fossil fuels to produce energy. The energy industry accounts for a large percentage of international greenhouse gas emissions, the response to this challenge needs to be a rapid and intensive shift to low-carbon sources of energy (Veers et al. 2022). Concurrently, energy demand is steadily increasing, driven by the expanding population, industrialization and urbanization, especially in the developing world. The growth in demand places heavy pressure on current energy infrastructure, making it imperative to have energy systems capable of delivering reliable and affordable energy to serve these needs in an efficient manner.

Another major challenge is energy security, which is impacted by geopolitical instability and weaknesses in global supply chains. The use of fossil fuels, especially oil and gas, has created energy dependencies that expose most countries to price volatility and supply shocks. Moreover, the shift to renewable energy sources such as solar, wind and hydropower also poses challenges. Although critical to sustainable energy options, renewable energy technologies are frequently confronted with challenges like variability in energy production, storage capacity and grid modernization needs. Finally, innovation in energy technologies, for example, energy storage technologies and grid management systems, is required to maximize energy systems and ensure efficient use of renewable resources (Hassan et al. 2024). These multifaceted challenges require creative solutions to energy planning and policy. Table 6.2 gives a concise and organized summary of the different world energy issues, their main problems and the solutions that can be used to solve them effectively.

6.1.2. *The role of energy system models in shaping the future*

Energy system models are sophisticated tools with a very active role to play in responding to the complex challenges of the global energy transition. They are a method of energy scenario modeling, and they enable policymakers and stakeholders to experiment and evaluate the likely effects of different strategies and technologies (Danish and Senjyu 2023). By simulating numerous alternative futures for energy, such models optimize the use of renewable energy in existing grids, reduce energy waste and keep energy supplies as steady in the long-run. Energy system models play a very significant role in forecasting future energy demand and examining how energy system issues like technology development, economic factors and policymaker choices will affect energy systems in the future (Anbalagan et al. 2024).

Area of impact	Example	Key metric	Value	Insight provided by models
Renewable energy integration	Wind and solar in national grid	Renewable share in 2040	65%	Assesses grid capacity to absorb renewable sources and identifies optimal investment paths.
Energy efficiency	Residential sector retrofit strategy	Energy savings by 2035	20–30% reduction in consumption	Helps estimate potential savings and cost-effectiveness of energy efficiency policies.
Carbon emission reductions	Introduction of carbon pricing	CO_2 emissions reduction by 2050	70% below 2020 levels	Models help simulate how taxes affect emissions under different price levels.
Economic impact of policies	Fossil fuel subsidy phase-out	GDP change by 2040	+1.2% (net economic benefit)	Shows macroeconomic impacts and helps guide equitable energy transitions.
Technology innovation impact	Widespread EV adoption	Oil demand reduction	30% by 2035	Simulates technological penetration and its impact on demand for fossil fuels.
Grid reliability and resilience	High renewable scenario with storage	System reliability index	99.95% uptime	Assesses risk of outages and needs for flexibility investments.
Long-term energy demand forecasting	Economic growth with population increases	Projected total energy demand in 2050	9,000 TWh	Projects demand growth and help plan for supply investments.
Climate policy scenario analysis	Net-zero 2050 policy simulation	Temperature rise limited	<1.8°C	Enables back casting from climate goals to required energy system transformations.

Table 6.3. *Simulating tomorrow: scenario planning in energy transformation*

These models are also planning and policy decision-support tools that offer helpful insights into the way governments can implement effective strategies to reduce greenhouse gas emissions and stimulate sustainable energy growth. Energy models can set out, through scenario analysis, how different policy interventions, such as carbon taxation, subsidization of renewables or energy efficiency standards, might influence energy systems and help to achieve climate targets (Cieplinski et al. 2021). In addition, energy system models enable long-term planning of energy to ready countries for future energy needs and disruptions so that more resilient and adaptable energy infrastructures can be developed. Table 6.3 indicates the key functions energy system models have in shaping future energy options through scenario analysis and strategic planning. It offers indicative numerical examples to demonstrate how economic models influence renewables, emissions, resilience and policy impacts decisions.

6.1.3. *Objectives and structure of the chapter*

This chapter explains the central contribution of energy system models toward a secure and sustainable world energy future. It illustrates different types of energy models, applications and their role in aiding global energy policy. This chapter revolves around how these models enhance the development of renewable energy, enhance energy efficiency, enhance energy security and meet climate objectives.

The chapter is organized as follows:

– *Section 6.2* gives an overview of the principal ideas and types of energy system models, including bottom-up, top-down and hybrid models, and their characteristics and uses.

– *Section 6.3* describes the application of these models to facilitating sustainable energy transitions with an emphasis on the way in which they facilitate the integration of renewable energy sources and improve overall system performance.

– *Section 6.4* outlines how energy options can be predicted by energy system models when it comes to the influence of technological innovation, economic transformation and policy intervention.

– *Section 6.5* looks at the significant contribution that energy system models can make to supporting government decision-making by providing tools for quantifying the performance of various policy options and aiding long-term energy planning.

– *Section 6.6* addresses the role of energy system models in cutting greenhouse gas emissions and attaining global climate targets, with emphasis on net-zero emissions.

– *Section 6.7* explains how energy models contribute to enhancing the resilience of world energy systems to handle possible threats and disruptions more effectively.

– *Section 6.8* concludes the chapter by summarizing the key findings and the potential of energy system models to address the challenge of a sustainable energy future.

6.2. Energy system models: key concepts and types

6.2.1. *Definition and scope of energy system models*

Aspect	Bottom-up models	Top-down models	Hybrid models
Modeling focus	Technological components and sectors	Macroeconomic variables and sectoral interactions	Integration of technological and economic modeling
Level of detail	High technological detail	Aggregated economic sectors	Combines detail from both approaches
Key features	Component-level analysis Technology-specific inputs Capacity expansion modeling	Macroeconomic representation Market-based responses General equilibrium potential	Dynamic feedback loops Tech-economic integration High modeling flexibility
Modeling approach	Bottom-up (from individual components to system behavior)	Top-down (from system-wide economic behavior to energy system outcomes)	Dual approach integrating micro (tech) and macro (econ) elements
Strengths	Detailed technology assessment Sector-specific insights Ideal for renewable energy planning	Broad economic insights Captures price effects and policy impacts Good for national-level decisions	Comprehensive perspective Captures both technological changes and economic effects Suitable for complex, systemic analysis
Applications	Energy supply and demand simulation Technology integration Sectoral decarbonization	Macroeconomic impact of policies Cost–benefit analysis Policy evaluation	Integrated climate-energy-economy modeling Comprehensive energy planning Scenario analysis for sustainability pathways
Typical users	Engineers, energy planners and sectoral analysts	Economists, policy analysts and government agencies	Interdisciplinary research teams, integrated policy developers

Table 6.4. *Comparative analysis of bottom-up, top-down and hybrid energy system models*

Energy system models are mathematical models that depict and simulate the behavior of energy systems and allow the user to investigate the effects of various variables and strategies on energy production, transmission, consumption and environmental conditions, as illustrated in Figure 6.2. Models are essential for decision-makers because they give insights into the possible effects of diverse policy choices, technological advances and economic changes. The range of energy system models varies from short-term operational modeling (e.g. balancing daily energy supply and demand) to long-term strategic planning (e.g. net-zero 2050). The main application of the models is to maximize the trade-off among energy efficiency, cost, security of supply and sustainability in light of the integration of renewable energy resources, energy storage and technology development. They also play an important role in determining the least expensive routes for attaining international climate and energy goals (Sridhar and Salkuti 2022).

6.2.2. *Bottom-up models: features and applications*

Bottom-up models are built to replicate energy systems by modeling individual components or sectors, such as power plants, buildings, transportation networks and industrial processes. Such models include elaborate technical details of energy technologies, including their efficiencies, costs and performance characteristics. Bottom-up models function by bringing together the performance of the components to mimic the behavior of the system (Fodstad et al. 2022).

Major characteristics of bottom-up models are as follows:

– *Technological detail*: they usually contain a very high degree of detail concerning the technologies applied in every industry, for example, energy production, storage and consumption technologies.

– *Component-level analysis*: these models break down individual energy components (e.g. solar panels, wind turbines and electric vehicles (EVs)) and examine their interactions within the system as a whole.

– *Capacity expansion*: bottom-up models are most effective to examine the capacity expansion of the renewable energy sources, the energy storage equipment and the efficiency measures with respect to time.

Applications of bottom-up models include the following:

– *Energy supply and demand simulation*: assessing the balance between energy supply and demand across different scenarios (e.g. integrating renewables and enhancing energy efficiency).

– *Technology integration*: evaluating the possibility of bringing new technologies (e.g. electric cars and smart grid management) into existing energy infrastructures.

– *Sectoral decarbonization*: how various sectors (e.g. electricity, transportation and industry) can decarbonize in alignment with climate targets.

6.2.3. *Top-down models: features and applications*

Unlike bottom-up models, top-down models depict energy systems in a less disaggregated form, highlighting macroeconomic factors and the aggregate interaction among various sectors of the economy. These models tend to use economic theory to represent the behavior of agents (e.g. firms, consumers and governments) and markets, capturing variables such as prices, income and demand elasticity. Top-down models incorporate top-level assumptions to represent the systemic impacts of energy policy and market forces on energy production and consumption (Zaiter et al. 2024).

Major characteristics of top-down models include the following:

– *Macroeconomic focus*: the models concentrate on macroeconomic variables such as GDP, investment and sectoral production.

– *Market-based interactions*: top-down models replicate the manner in which energy markets react to policies, technology prices and price signals (e.g. carbon pricing and subsidies).

– *General equilibrium*: they sometimes use a general equilibrium model, describing how a change in one component of the economy influences the rest of the system.

Applications of top-down models include the following:

– *Macroeconomic impact assessment*: it includes assessing the general economic effects of energy policies, for example, carbon taxes or renewable subsidies.

– *Cost–benefit analysis*: it includes evaluating the economic trade-offs of various energy policies or technological developments.

– *Policy evaluation:* it includes examining the effectiveness and economic impact of climate policies at a sectorial level.

6.2.4. *Hybrid models: a combined approach*

Hybrid models integrate the strengths of both bottom-up and top-down models, using the best of each to offer a more complete picture of energy systems. Hybrid models seek to incorporate the detailed technological details of bottom-up models and the macroeconomic interactions of top-down models.

Hybrid approaches are especially valuable when both technological detail and economic interactions are at issue, as with the study of large-scale energy transformations or the integration of renewable energy technologies into current economic institutions (Shoaei et al. 2024).

Major characteristics of hybrid models are as follows:

– *Technological and economic integration*: hybrid approaches put together sectoral or technological modeling in detail with macroeconomic modeling, enabling a more integrated analysis.

– *Dynamic interactions*: they facilitate the analysis of feedback loops between technological advances, economic growth and policy action.

– *Flexibility:* the models can be tailored to fit a large variety of situations, ranging from energy system optimization, impacts of climate policy and future technology costs.

Applications of hybrid models include the following:

– *Comprehensive energy system planning*: assessing the costs and advantages of shifting to low-carbon energy systems taking into account technological, economic and policy considerations.

– *Integrated climate-energy-economy modeling*: understanding how energy, economic and climate policies impact each other for long-term climatic objectives.

– *Scenario analysis*: evaluating various policy, technology and economic scenarios to determine routes to a sustainable energy future.

Table 6.4 compares bottom-up, top-down and hybrid energy system models side-by-side in terms of their fundamental features, modeling practices and applications. It offers a quick overview of how each model type enables energy planning as well as policy analysis.

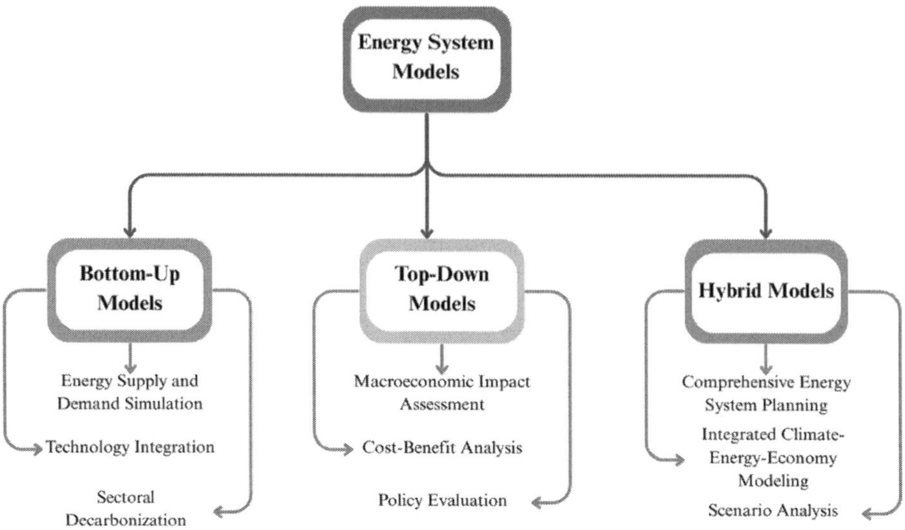

Figure 6.2. *Classification and applications of energy system models. For a color version of this figure, see www.iste.co.uk/sundaram/energysystems.zip*

6.2.5. *Model selection criteria and best practices*

Choosing an appropriate energy system model relies on a number of factors, ranging from the purpose of the research to the degree of detail needed, as well as the nature of decision-making involved (Gupta et al. 2025).

The main selection criteria include the following:

– *Scope and purpose*: the model should be tailored to the specific objectives of the analysis, whether that is analyzing short-term energy reliability, long-term decarbonization or economic effects of energy policies.

– *Data availability*: quality and availability of data for energy systems, technology prices and economic factors will determine the selection of models. Some models need very detailed sectoral data, whereas others are concerned with macroeconomic aggregates.

– *Model complexity*: complex models (e.g. hybrid models) can offer a close-up examination but might be data and computationally intensive. Basic models are probably suitable for first-cut scoping or policy appraisal.

– *User expertise*: another key factor is the expertise of the modeler. There might be advanced technical knowledge required in some models related to energy systems or economics, whereas some could be relatively simple for policymakers and stakeholders to access.

Best practices in model use include the following:

– *Transparency and documentation*: documentation of the assumptions, methodologies and data sources of the models so that they are transparent to build confidence in the results.

– *Validation and calibration*: periodic validation and calibration of the model with actual data to enhance its predictability and credibility.

– *Sensitivity analysis*: performing sensitivity analyses to determine the impact of changes in key assumptions or parameters on the model results to understand the robustness of the results.

– *Stakeholder engagement*: engaging stakeholders early in the modeling process to guarantee that the model captures real-world constraints, goals and uncertainties.

6.2.6. *Comparison of existing energy system models*

Current energy system models differ in strategy: bottom-up models concentrate on technology and optimization of the system, top-down models focus on economics and market behavior and hybrid models are a mix of both. Bottom-up models are excellent at predicting technologies, top-down models provide economic data and hybrid models attempt to do both. The model chosen is based on the particular aims, data availability and level of detail required for the analysis (Behrens et al. 2024).

Energy system models may be classified into three categories: bottom-up, top-down and hybrid. Bottom-up models such as TIMES, MARKAL and LEAP are technically oriented, illustrating how various technologies (such as solar panels or EVs) can be applied to satisfy energy demands at least cost. They are excellent tools for investigating clean energy options. Top-down models like GEM-E3, CGE, PEW and MESSAGE take a step back by connecting the economy with energy use. They assist in illustrating how energy policy impacts jobs, prices and trade. Hybrid models, such as EnergyPLAN and E3ME, contain comprehensive technology data and illustrate how they impact the economy. These are particularly useful for the examination of complicated policies, such as those concerning climate change or national energy planning. Table 6.5 offers a comparative summary, encompassing both quantitative information and qualitative conclusions. Comparison should assist in comprehending the capabilities and limitations of each model type in various energy system analysis situations.

Aspect	Bottom-up models	Top-down models	Hybrid models
Example models	TIMES, MARKAL, LEAP	GEM-E3, CGE, PEW, MESSAGE	EnergyPLAN, E3ME
Computational time (average)	12–48 h for a regional model (depends on complexity)	5–20 h for regional models	24–72 h (depending on hybrid complexity)
Data requirements	10–100 GB of detailed technological data (for a regional system)	2–10 GB for economic and macroeconomic data (annual datasets)	50–150 GB (mix of detailed and economic data)
Number of scenarios	50–500 scenarios per run (depending on system complexity)	10–100 scenarios per run	20–200 scenarios per run
Accuracy of technological forecasts	High (technology-specific forecasts)	Moderate (macroeconomic aggregates, less focus on technology specifics)	High (combines both detailed tech and economic forecasts)
Energy system sectors modeled	Focus on energy technologies (e.g. electricity and heating)	Focus on macroeconomics (e.g. GDP and demand)	Comprehensive coverage: energy, transport, industry and so on
Sectoral integration	Limited (mostly electricity or heating)	High (integrates economy-wide sectors)	High (integrates multiple sectors, but complexity increases)
Data resolution	High (technology and sector-specific data)	Low (aggregate data on sectors, economy)	Intermediate (technology and macroeconomic data)
Model resolution (granularity)	High (individual energy technologies and infrastructure)	Low (economy-wide interactions)	Intermediate (blends technology-specific and economic-level analysis)
Flexibility	Low (more rigid structure due to detailed tech focus)	High (easier to modify macroeconomic assumptions)	Moderate (trade-off between technology detail and economic assumptions)
Real-time adaptability	Low (difficult to update once scenarios are set)	Moderate (can update economic assumptions)	Moderate (updates required for both sectors and technologies)
Cost of data acquisition	High (requires detailed, high-quality tech data)	Moderate (rely on publicly available economic data)	High (requires both technical and economic data sets)
Limitations	Poor at modeling complex economic feedbacks or unexpected shifts	Less precise in modeling specific technological innovations	Computationally intensive, complexity in balancing both models

Table 6.5. *Comparison of existing energy system models: limitations, development opportunities and statistical data*

6.2.7. *Limitations of existing energy system models and scope for development*

Limitation	Description	Scope for development	Potential solutions	Statistical data	Model accuracy impact
Limited representation of technological innovation	Existing models often fail to incorporate disruptive or rapidly evolving technologies (e.g. energy storage, CCS, smart grids).	**Incorporating innovation and technology forecasts**	Integrate learning curves and model technology adoption dynamics.	*Computational time*: 5–10% increase due to additional technology scenarios. *Data requirements*: +20–30 GB.	+10–20% **(better long-term projections of emerging tech adoption)**
Inadequate modeling of behavioral and policy dynamics	Models tend to oversimplify how policies and societal behaviors interact with energy transitions (e.g. market dynamics, public acceptance).	**Integrating behavioral and policy dynamics**	Develop agent-based models and policy feedback loops.	*Computational time*: +10–15% due to more dynamic, interactive policy simulations. *Data requirements*: +10 GB.	*+10–15%* **(improves realism in policy and social behavior outcomes)**
Data quality and availability	Inaccurate, outdated or incomplete data, especially in developing regions, limits the effectiveness of models.	**Improving data acquisition and standardization**	Establish global datasets, enhance real-time data integration.	*Data requirements*: +50–100 GB for better data integration. *Resolution*: Low to high depending on data quality.	*+20–30%* **(significant improvement, especially in data-poor regions)**
Complexity and usability	Many models are highly technical and difficult for non-experts to use, limiting their effectiveness for decision-makers.	**Simplification and user-friendly tools**	Create simplified user interfaces and decision-support tools.	*Computational time*: +10% for simplified model versions. *User **accessibility***: +50% improved with new interfaces.	*±0%* **(no direct accuracy change; improves usability rather than model quality)**
Limited integration across sectors	Many existing models focus on a single sector (e.g. electricity) and do not integrate other energy sectors like transport or heating.	**Cross-sector integration**	Build multi-sectoral models that combine electricity, transport, heating and industrial energy systems.	*Computational time*: +20–40% due to increased complexity. *Data **requirements***: +100–150 GB for cross-sector data.	*+15–25%* **(provides a more complete system-level picture)**

Uncertainty and risk analysis	Most models fail to account for uncertainty in key variables such as technology costs, policy changes and geopolitical events.	**Uncertainty and sensitivity analysis**	Use probabilistic models and Monte Carlo simulations to incorporate uncertainties.	*Computational **time***: +15–30% for Monte Carlo simulations. *Accuracy*: +10–20% improved with uncertainty integration.	*+10–20%* **(more reliable and robust decision-making under uncertainty)**
Real-time adaptability	Current models are typically static and cannot adjust in real-time to new information or unexpected events (e.g. energy crises, policy changes).	**Real-time dynamic modeling**	Implement adaptive, machine learning-based models that can adjust in real time.	*Computational time*: +30–50% for real-time data updates. *Data requirements*: +200 GB for continuous data updates.	*+15–25%* **(faster adaptation to unexpected events improves relevance)**
High cost of data acquisition	Data required for running high-resolution energy system models can be expensive and difficult to obtain, especially for developing regions.	**Cost-effective data collection**	Leverage open-source databases, collaborate with international organizations to reduce costs.	*Data costs*: $50,000–$200,000 per model depending on resolution and scope of data required.	*+5–15%* **(depends on ability to substitute high-cost data effectively)**
Limited regional and local modeling	Most models focus on global or national levels and fail to account for regional or local energy dynamics (e.g. local energy resources, grid stability).	**Regional and localized modeling**	Develop high-resolution regional models that incorporate local energy characteristics and resources.	*Computational **time***: +20–40% for regional simulations. *Data **requirements***: +50–100 GB for regional specifics.	*+10–20%* **(better spatial accuracy and relevance for local decisions)**
Lack of transparency in model assumptions	Many models use black-box methodologies, making it difficult for users to understand how assumptions and inputs impact the results.	**Increased transparency in model assumptions**	Improve model transparency with clearer documentation and open-source platforms.	*User accessibility*: +25% improved with open-source models. *Computational **time***: +10% for assumption validation.	*+5–10%* **(increases trust and interpretability enabling better-informed decisions)**

Table 6.6. *Limitations of current energy system models and pathways for development: statistical insights*

The shortcomings of current energy system models are attributed to a number of factors, such as their inability to accurately capture technological innovations, intricate policy interactions and adaptability to real-time data. Most models also have challenges with high data demands, sectoral integration and making accurate predictions under uncertainty. The areas of development include improving model flexibility, enhancing data quality, integrating multi-sectoral dynamics and incorporating real-time adjustments (Denysov et al. 2024). Improvements in computing power, machine learning and international data cooperation present good opportunities to overcome these obstacles, providing more accurate and useful information for energy planning and policy decision-making.

Table 6.6 presents an organized summary of the restrictions and development potential for energy system models, presenting definite statistics regarding how progress might impact model performance and resource demands.

6.3. The role of energy system models in sustainable energy transitions

6.3.1. *Promoting renewable energy integration*

One of the most important functions that energy system models have in the transition to sustainable energy is encouraging the incorporation of renewable energy sources into the current energy systems. As worldwide efforts to curtail greenhouse gas emissions gain momentum, there is a growing urge to substitute fossil fuels with renewable energy sources such as wind, solar, hydro and geothermal. But integrating such variable sources is difficult, with major challenges being the stability of the system, energy storage and optimizing the operation of the grid. Energy system models provide a solution for these challenges because they simulate how renewable energy technologies behave in a larger energy system (Cedeno and Wei 2024). They therefore give insights about the optimal mix of renewable sources and assist in designing energy infrastructure that can cope with such changing sources. Using scenarios with varied levels of penetration of renewable energies, these models can identify possible costs, benefits and feasibility of integrating these renewables into the grid. Moreover, they can study how the improvement in storage technologies (e.g. battery storage and pumped hydro) and increased flexibility of the grid can help manage the intermittency of renewable energy generation for a smooth flow of electricity supply.

6.3.2. *Enhancing energy efficiency*

Energy system models are also crucial for increasing energy efficiency in the sectors of electricity generation, transportation and industry. Improving efficiency is an important element of moving toward sustainable energy systems because it lowers the total energy consumption, reduces expenditure and lessens environmental burdens. Energy system models allow efficiency gains to be modeled and allow policymakers to be aware of the impact on the system (Krishna et al. 2024). These models can assess a broad variety of energy efficiency opportunities ranging from retrofitting the existing infrastructure and installing energy-saving technology to reducing home and business energy consumption. Through an analysis of the least-cost ways of making energy savings, energy system models assist decision-makers in ranking investments in energy efficiency by industry. They can also forecast long-term efficiency gains in terms of decreased energy consumption, decreased emissions and lower dependence on non-renewable fuels. This can help to inform policy and strategy development that is capable of supporting energy transition and additional contributions toward the overall goals of sustainability and decarbonization.

6.3.3. *Ensuring security of energy supply*

In the context of an energy transition globally, energy security ranks as a priority. Energy system models are significant in guaranteeing that there is a steady and constant supply of energy, particularly as nations move away from fossil fuels to cleaner, more decentralized sources of energy. Introducing renewable energy can be difficult in terms of the supply reliability and consistency to the extent that some renewable resources such as wind and solar are intermittent and variable. Models of energy systems can be used to determine the energy supply to determine the security of energy supply by analyzing the various energy scenarios and establishing areas of vulnerability to speculative assaults (Das et al. 2024). Models can examine the susceptibility of energy infrastructure to disruption by extreme weather patterns, geopolitical tensions or natural disasters that can impact energy production and supply. Models can also determine the impact of different energy storage devices and grid enhancement on supply reliability (Emrani and Berrada 2024). By foreseeing probable supply-demand mismatches, these models enable governments and energy suppliers to plan strategies for guaranteeing energy supply stability, either through increased grid interconnections, standby power arrangements, or multiple sources of power. Models of the energy system thus help ensure a balancing act between ensuring security and meeting demands for decarbonization so that transitions away from carbon-intense power supplies do not negatively impact energy reliability. Some energy models assist nations to ensure that they get a safe and reliable supply of energy. For instance, EnergyPLAN indicates how well the energy system of a country can cope with issues such as fuel shortage or price fluctuations. It assists in planning for more use of local energy and less from abroad. MESSAGE considers long-term energy strategies and assists in identifying means to prevent overdependence on a single form of energy or provider. NEMS, employed in the United States, verifies how energy policy impacts domestic energy supply and imports. The models assist governments in identifying and correcting vulnerabilities in their energy systems and ensuring that energy is secure for the future.

6.3.4. *Addressing environmental and geopolitical pressures*

Dimension	Role of energy system models	Key impacts	Benefits for decision-makers	Examples of relevant models
Environmental	Forecast impacts of energy policies on GHG emissions and air quality Simulate effects of carbon pricing, ETS, renewable incentives	Quantify emissions reductions Highlight environmental trade-offs Assess sustainability impacts	Align policies with climate targets Design environmentally responsible strategies	LEAP (Low Emissions Analysis Platform) GCAM (Global Change Assessment Model)
Geopolitical	Analyze global energy transition impacts on trade, security and energy dependencies	Reveal new dependencies (e.g. rare earths) Forecast market shifts Identify emerging energy powers	Enhance national energy security Anticipate geopolitical tensions or opportunities	NEMS (National Energy Modeling System) POLES (Prospective Outlook on Long-term Energy Systems)
Economic equity	Evaluate effects of transition on vulnerable populations and fossil fuel-dependent regions	Identify socio-economic disparities Highlight risks of regional decline or unrest	Support a just and inclusive transition Prioritize support for affected communities and regions	TIMES (The Integrated MARKAL-EFOM System) REMIND (Regional Model of Investment and Development)
Strategic planning	Integrate both environmental and geopolitical insights for long-term resilience	Synthesize cross-sectoral impacts Provide holistic transition pathways	Enable balanced, informed policymaking Foster global stability and sustainable development	MESSAGE (Model for Energy Supply Strategy Alternatives and their General Environmental Impact) OSeMOSYS (Open-Source energy Modelling System)

Table 6.7. *Integrating environmental and geopolitical considerations into energy system modeling*

Energy system models also have a key role to play in responding to the wider environmental and geopolitical forces that influence energy policy and choice. Environmentally, models assist in predicting the effect of energy policy on greenhouse gas emissions, air quality and other environmental results. By simulating the effects of various policy interventions, such as carbon prices, emissions trading schemes or renewable energy incentives, models enable governments and institutions to develop strategies that are aligned with international climate and sustainability goals. Geopolitically, energy system models help evaluate the consequences of global energy transitions on energy security, trade relations and energy dependencies (Bakhsh et al. 2024).

As nations move toward renewable energy, some long-time fossil fuel exporters could be economically strained or find benefits in emerging energy markets. Energy models can project the potential geopolitical impacts of these transitions so that policymakers can observe how changes in world energy demand and supply might impact international relations, market prices and energy security.

Additionally, energy system models can help tackle the distributional effects of the energy transition to ensure that the transition to renewable energy is equitable and just. This entails a review of impact on vulnerable countries, fossil fuel-dependent nations and most vulnerable nations to energy price shocks (Wu and Hussain 2025). Both geopolitical and environmental analysis are enabled through energy system models, helping decision-makers to make well-balanced and well-informed choices that support global stability as well as sustainability. There are some energy models that are intended to demonstrate the implications of energy consumption on the environment, particularly the emission of greenhouse gases and pollutants.

For instance, LEAP is often utilized to monitor emissions from various energy sources and simulate whether clean energy policy can curb pollution. GCAM examines the interactions between energy, land use and climate, assisting in the long-term environmental effects of energy decisions. TIMES also has comprehensive emissions data and assists in investigating how a transition to cleaner technologies can cut carbon emissions. The models assist policymakers in designing energy systems that accommodate climate objectives and conserve the environment.

Table 6.7 provides a concise overview of the role energy system models can play in the reaction to environmental and geopolitical challenges in the field of energy policy. It highlights the ability of the models to forecast the environmental impacts, such as greenhouse gases and air pollution, and examine geopolitical elements like energy security, trade shifts and global interdependencies.

6.4. Predicting future energy scenarios

6.4.1. *Projections of renewable energy growth*

Forecasting future growth in renewable energy sources is one of the key functions of energy system models. With the world collectively working to reduce dependence on fossil fuels and lower greenhouse gas emissions, renewable energy will probably have an increasingly important role in the global energy mix. Energy models are capable of making predictions of the expansion of renewable energy technologies, that is, solar, wind, hydro and bioenergy over time based on drivers like technological developments, policy initiatives and market trends (Joshi et al. 2025). Renewable energy growth projections usually entail the simulation of different scenarios with different degrees of technological advancement, policy action and market conditions. These models take into account declining renewable technology costs, energy storage innovation and the integration of renewables into grids that can handle variability. For example, with the swift decreases in the price of solar and wind energy, energy system models can estimate the way these resources will increasingly take the place of fossil fuels as electricity sources. Models can also predict the size of the chance for renewable energy to provide growing shares of worldwide energy consumption in the mid-21st century, guiding investments and policy (Modis 2022). Major takeaways from these forecasts are to recognize the places and industries that renewable power will expand most rapidly, comprehension of grid integration and infrastructure issues, and demarcation of the place of energy storage and demand management in securing a stable supply.

6.4.2. *Impact of technological breakthroughs on energy systems*

Technological innovations hold the potential to transform energy systems by making energy production, storage and use more efficient, economical and sustainable. Energy system models are key to projecting the impact of these innovations and determining how these could transform the energy sector. Energy technology innovations, including improvements in energy storage, carbon capture and storage (CCS) and next-generation nuclear energy, have the potential to transform future energy circumstances. For instance, advances in battery technology have the potential to accelerate the uptake of EVs and grid-scale storage technologies, mitigating the intermittency issues related to renewable energy generation (Ahmadizadeh et al. 2024). In the same way, advances in smart grid technologies could improve the flexibility and resilience of energy systems and facilitate more efficient management of renewable energy sources. Energy system models can model the effects of such technological advances by incorporating them into different energy scenarios. They assist in projecting how these innovations will impact energy prices, efficiency, emissions savings and the rate at which renewable energy can be ramped up. By predicting the likely contribution of these technologies, energy system models allow policymakers, investors and companies to make better decisions on future energy infrastructure and development.

6.4.3. *Economic trends and their influence on energy models*

Economic trends are major drivers of the future of energy systems. Since energy models depend on macroeconomic factors like GDP growth, energy prices and investment trends, economic trends must be known for making reliable long-term energy demand and supply forecasts. Economic trends affect both the energy demand (e.g. industrial development or urbanization) and the cost-effectiveness of energy technologies (e.g. investment in renewable energy and infrastructure). Energy system models incorporate economic factors such as market trends, economic growth and the decline in technology costs to simulate how the economy will influence energy consumption and production (Dharmapriya et al. 2024). For instance, economic growth in developing economies is likely to drive rapid growth in energy demand, which consequently affects the energy mix. Economic downturns or recessions, in contrast, will reduce energy demand, which affects energy infrastructure investment volumes. Energy models also reflect the effect of economic policies such as carbon pricing or subsidies for renewables on energy market dynamics. Such economic incentives can accelerate or decelerate the shift toward green energy systems through modifications in the cost-competitiveness of renewables vis-a-vis fossil fuels. Economic factors also influence the pace of adopting technology, that is, strong economic recovery can be a measure of ramped-up investment in green technology, but economic uncertainty can hinder the pace of implementing infrastructure projects.

6.5. Energy system models and policy making

6.5.1. *Supporting policy decisions for energy transition*

Energy system models play an essential role in informing policy action for the energy transition to be successful. With governments everywhere setting ambitious reductions in greenhouse gas emissions, increasing the proportion of renewables and setting net-zero emission targets, energy system models provide essential intelligence on how best to do that. By simulating the interactions between energy supply, demand and infrastructure, these models allow policymakers to understand the probable impacts of different policy interventions and choices. For instance, models can assess the impact of subsidies for renewable energy, carbon taxes or mandates on the energy market (Alamoush et al. 2024). They make it possible to identify the most effective policies for accelerating the deployment of renewable energy and improving energy efficiency. Energy system models also have the ability to determine how the above policies will affect various sectors, areas and socioeconomic groups, such that energy transition strategies are fair and inclusive. In addition, energy models support policymakers in analyzing the possible drawbacks and trade-offs in the process of energy transformation, for example, reconciling economic development and decarbonization or coping with social and economic implications of an energy transition out of fossil fuels (Heras and Gupta 2023). By indicating with evidence-based foresight projections, which policy options will affect the energy systems how, these models are key policy-making tools that respond to the imperatives both of short-run and long-run sustainability.

6.5.2. *Evaluating the effectiveness of policy measures*

Once implemented, energy policies must be analyzed to determine whether they are producing the desired results. Energy system models are crucial to this process by providing a framework to examine the effects of policy interventions on energy systems in the long-term. These models are able to reproduce the actual real-world effect of policies, such as carbon prices, renewable energy targets, or energy efficiency measures, and monitor their progress in reaching established objectives. By contrast between various policy scenarios, energy models assist in identifying whether available measures are successful in curbing emissions, accelerating the adoption of renewable energy, or enhancing energy security. Energy models can further identify where policies may need to be adjusted or improved to meet energy transition goals (Ekechukwu and Simpa 2024). Energy models of systems may also have the capacity to forecast unintended consequences of policy decisions. For example, while a carbon tax would stimulate the application of clean technologies, it might increase the price of energy for low-income families. Energy sector models can evaluate such potential trade-offs and help policymakers design compensatory schemes that will be fair and avoid negative social or economic consequences. Overall, energy system models provide the analysis necessary to evaluate the impacts of policies and improve them so that they can effectively guide the transition to sustainable, low-carbon energy systems.

6.5.3. *Energy models in long-term planning and scenario analysis*

Long-term planning is needed to provide for future demand while transitioning to low-carbon, sustainable forms of energy. Energy system models are crucial to long-term energy planning since they allow for the exploration of various future scenarios under a wide range of uncertainties and external drivers, such as technological advancements, economic conditions, and environmental policy. Scenario analysis is one of the methodologies through which energy models explore alternative paths of the energy transition considering the pace of renewable energy integration, the path of developing new technologies (e.g. hydrogen or energy storage), and implications of global trends (e.g. climate shocks or geopolitical transformations). Scenario models allow stakeholders to envision potential future outcomes of varying decisions and gear up for an energy future that is secure, sustainable and resilient (Nicoli et al. 2024). By simulating different situations, energy models also allow planners to identify points of critical decision-making, that is, when to invest in new infrastructure or when specific technologies will be commercially available. Long-term planning with energy models provides the foresight to predict future uncertainties and challenges and ensure energy systems evolve in a way that promotes global sustainability goals, that is, keeping global warming well below 2°C. Moreover, energy system models are valuable tools for synthesizing the perspectives of various stakeholders in long-term energy planning. They can lead discussions among governments, energy companies, environmental organizations and the public through providing an open and transparent framework for understanding likely outcomes and trade-offs.

6.6. Impact of energy system models on greenhouse gas emissions and climate goals

6.6.1. *Modeling the path to net-zero emissions*

Mid-century net-zero emissions have become among the high-priority goals of most countries and organizations globally owing to pressure from the top on combating climate change. The models of the energy system are confronted with replicating the net-zero emissions trajectory through an examination of how different sectors such as electricity generation, transportation, industry and buildings can reduce the carbon intensity (Kaplan et al. 2025). Through the duplication of the effects of different policy interventions, technological developments and changes in consumer tastes, such models enable policymakers to have a clear and feasible roadmap toward achieving the net-zero aspirations.

Planners are able to measure the most efficient and technologically suitable decarbonization options through the assistance of energy system models. For example, the models can indicate the way in which renewable energy supplies such as solar, wind and hydropower can supplant fossil fuel use in generating electricity, or the way transportation electrification can minimize oil use. They also examine other alternatives including CCS, measures of energy efficiency and lifestyle changes that can further reduce emissions in sectors. Energy models give in-depth information about the timing, investment and infrastructure needed for a transition to net-zero emissions that will be successful (Ross-Hopley et al. 2025). They can establish significant milestones, including when main technologies (e.g. green hydrogen, next-generation nuclear or EVs) will become commercially available and have broad adoption. The models are fundamental tools to ensure that global and national climate goals are achieved in a sustainable and economically justifiable manner. Table 6.8 provides a sector-by-sector statistical framework that maps out the pathway toward 2050 net-zero emissions. It displays existing and expected CO_2 emissions for every significant sector such as power generation, transport, industry, buildings, agriculture and carbon removal along with the main decarbonization methods, key technology likely to grow to mid-century and the estimated 2020–2050 investment required. The sequential format places strong emphasis on the central role of energy system models in coordinating sustainable, economically sensible transitions between sectors to achieve global climate goals.

Sector	2020 CO_2 emissions (Gt CO_2)	2050 Target emissions (Gt CO_2)	Primary decarbonization strategy	Key technology deployment by 2050	Cumulative investment needed (2020–2050, USD trillion)
Power generation	13.5	~0.2	Full shift to renewable energy and nuclear	98% renewables + advanced nuclear, CCS	15.0
Transportation	7.2	~0.3	Electrification, hydrogen fuels, modal shift	100% EV adoption, green hydrogen in freight	8.5
Industry	6.9	~1.0	Electrification, CCS, green hydrogen, efficiency	50% hydrogen use, CCS for cement/steel	10.0
Buildings	3.0	~0.1	Electrification, heat pumps, efficiency improvements	90% heat pump adoption, low-energy buildings	4.0
Agriculture and land	6.0	~1.5	Sustainable practices, methane reduction, reforestation	25% methane reduction, net afforestation	3.0
Carbon removal	–	–3.1 (net negative)	Afforestation, DACCS (Direct Air Carbon Capture and Storage), BECCS (Bioenergy with Carbon Capture and Storage)	Scaled deployment of DACCS and BECCS	2.5

Table 6.8. *Sectoral roadmap to net-zero emissions by 2050: statistical outlook and investment requirements*

6.6.2. *Role in achieving global climate objectives*

Energy system models are important in assisting nations to reconcile their energy policies with international climate targets, for instance, the targets in the Paris Agreement. To cap global warming at well below 2°C and preferably 1.5°C, nations must collaborate to transform their energy systems to be low carbon. Energy system models assist in determining what contribution each nation or region can make, depending on its specific energy resources, economic situation and requirements. They also demonstrate the best mix of policies, technologies and approaches for reducing emissions globally (Evro et al. 2024). For instance, models are able to determine the way higher shares of renewables, energy efficiency and utilization of carbon markets can improve climate goals. They also illustrate the way worldwide cooperation, such as clean technology transfer or extending finances to developing nations, can accelerate low-carbon planetary transformation. By exploring the long-term impact of different climate policy decisions, energy models offer policymakers the insight they need to create feasible and ambitious climate policies that are consistent with global ambitions.

6.6.3. *Case studies and success stories*

Actual case studies and success stories show where energy system models have been applied successfully in developing successful energy transitions and greenhouse gas mitigation targets. Regions and nations have applied energy system models to utilize policy successfully that includes renewable energy, raises the level of energy efficiency and decreases carbon emissions.

A good case in point is how Denmark used energy models in transitioning to a near-renewable electricity system. Denmark established aggressive climate targets and used energy system models to help guide the integration of wind power and prepare the grid for renewable energy. These models enabled the determination of the best energy storage, grid and demand response system investments, resulting in an unprecedented proportion of renewables as part of the country's electricity mix. Denmark is today regarded as a global model for the deployment of renewables, especially wind power and energy models were crucial in facilitating the transition.

A second example is that of the United Kingdom, which employed models of the energy system to create the carbon budgets established by the committee on climate change (CCC). The CCC employed models to chart the trajectory for the UK to meet its legally binding carbon reduction targets and assist in establishing the policies to be implemented to realize the targets without sacrificing economic growth. Consequently, the UK has observed a significant decrease in power generation emissions, primarily because of the transition from coal to natural gas and renewable sources like wind and sunlight.

Category	Denmark	United Kingdom	California (USA)
Primary objective	Transition to a renewable-based electricity system	Meet legally binding carbon budgets and reduce national GHG emissions	Achieve 100% clean electricity by 2045 and advance clean energy innovation
Key energy transition strategy	Integrate wind power; optimize grid and storage	Transition from coal to gas and renewables; maintain economic growth	Decarbonize electricity, support EVs, expand solar and storage technologies
Role of energy system models	Guide grid optimization, renewable integration, storage and demand response investments	Develop carbon budgets; identify cost-effective emission reduction pathways	Explore decarbonization scenarios; inform policy for EVs, solar adoption and energy storage
Main outcomes	High share of renewables in power mix; global leader in wind energy	Significant drop in power sector emissions; strong policy foundation for climate action	Leadership in U.S. clean energy policy; rapid adoption of solar, EVs and clean tech
Policy impact	Supported national climate targets and investment strategies	Informed strategic policy frameworks by the committee on climate change (CCC)	Enabled ambitious legislation (e.g. 100% clean energy law); steered technology and infrastructure growth
Global recognition	Model for renewable energy integration, especially wind	Recognized for pioneering carbon budgeting and emission reductions	Seen as a clean energy policy innovator in the United States
Energy model functions	Scenario analysis, investment planning, integration of renewables with existing infrastructure	Pathway modeling, economic impact analysis, policy assessment	System decarbonization modeling, impact evaluation, long-term strategic planning

Table 6.9. *Comparative case studies: real-world applications of energy system models driving energy transitions and reducing greenhouse gas emissions*

In California, models of the energy system have supported the State to achieve its ambitious clean energy and climate goals, such as policy mandating 100% clean electricity by the year 2045. In modeling alternative trajectories for decarbonizing its energy system, California has made policy decisions to facilitate the growth of EV usage, solar and the promotion of energy storage technologies. They have helped California become the premier energy innovating and climate regulating state in America.

These case studies indicate that energy system models not just provide important insights to achieve emission reduction objectives but also enable efficient climate policies to be enacted. Their successful application demonstrates how effective they are in informing policy decisions and facilitating mass-scale energy transitions.

Table 6.9 demonstrates how the models have been applied successfully in Denmark, the United Kingdom and California to facilitate energy transitions and reduce greenhouse gas emissions. Highlighting the central role of the modeling tools in defining sustainable and cost-effective energy options, the table focuses on the major goals, strategies, outcomes and impacts of the policies.

6.7. Energy system models and global energy resilience

6.7.1. *Strategies for energy resilience in a changing world*

Against the backdrop of rising challenges like climate change, geopolitical rivalries and rising energy needs, nations, sectors and organizations globally are setting energy resilience as a priority. Energy resilience is a situation where energy systems can foresee, prepare for and respond to disruptions and continue with normal and safe energy supply. As energy systems become increasingly complex and interconnected, energy system models play a crucial role in formulating strategies that maximize resilience. Modeling of energy systems assists in identifying the vulnerabilities of energy infrastructure and supply chains like dependence on fossil fuels, out-of-date infrastructure or the consequences of climate-induced events like storms, flooding and drought. Through simulation of different scenarios such as extreme climate events, conflict or failure of technology, models assist the policymakers in understanding the potential effect of these perturbations to energy networks. Models can also ascertain the feasibility of various resilience measures, such as diversification of energy supply, grid flexibility, investment in energy storage and construction of more resilient infrastructure. Moreover, energy system models can facilitate planning for more decentralized energy systems, which could be more resilient during crises. For example, decentralized renewable energy sources like rooftop solar power and local batteries can ensure a continuous supply of energy in the event of grid failure. Through imitating these measures, energy system models allow the creation of policy and investment that makes energy systems resilient in a changing and volatile world.

6.7.2. *The role of energy models in crisis management*

Energy system models are also valuable tools in crisis management, particularly in instances where unforeseen disruptions, for example, natural catastrophes, economic shock, or geopolitical turmoil, impact energy systems. Models allow for fast, data-informed decisions to mitigate the effect of energy crises and facilitate restoring energy systems to full function.

Crisis type	Use of energy system models	Example	Impact
Natural disasters	Forecast supply disruptions, simulate grid damage, optimize recovery pathways	Japan (2011 Tōhoku Earthquake) – modeled the energy shortfall after nuclear shutdowns	Helped reallocate generation, supported short-term fuel switching, informed recovery of energy mix
Geopolitical conflicts	Simulate fuel supply risks, evaluate import dependencies, model alternative scenarios	Europe (2022 Russia-Ukraine war) – modeled gas import disruptions and accelerated renewable scenarios	Informed decisions on gas storage, LNG import infrastructure and RE policy acceleration
Economic shocks	Assess demand fluctuations, optimize system operation under reduced consumption	COVID-19 Pandemic (Global) – modeled shifts in energy demand and impact on power markets	Supported flexible grid operation, deferred investments, guided economic recovery planning
Infrastructure failure	Analyze cascading failures, identify critical nodes, plan system reconfiguration	Texas Power Crisis (2021) – modeled extreme weather stress on grid and generation systems	Highlighted system vulnerabilities, prompted winterization policies, updated capacity planning models
Fuel supply interruptions	Optimize fuel allocation, simulate supply chain recovery, assess substitution possibilities	Hurricane Katrina (2005) – modeled oil and gas supply disruption in the U.S. Gulf Coast	Supported emergency fuel logistics, informed DOE response, highlighted resilience needs
Post-crisis recovery	Evaluate rebuild timelines, estimate cost–benefit of resilience investments	Puerto Rico (Post-Hurricane Maria) – used models to redesign a resilient, decentralized energy system	Guided investments in microgrids and renewables, improved system robustness

Table 6.10. *Energy system models in crisis scenarios: applications across different types of disruptions*

During crisis conditions, energy models can forecast the short-term effect of disruptions on energy demand, production and supply, allowing decision-makers to better understand the optimal way to allocate resources and control energy reserves. For example, models can forecast the effect of a natural disaster on oil or gas production and decide whether alternative energy sources, such as renewables or standby systems, can meet demand. They can also examine the potential benefits to be derived from demand-side management policies, such as rolling blackouts or energy conservation campaigns, to reduce the load on the power system in periods of severe crisis. Besides, energy system models enable analyzing the recovery and reconstruction process following a crisis (Fotopoulou et al. 2023). Energy models can replicate the time to restore from damaged facilities, the financial efficiency of the repair method and the most efficient distribution of resources in rehabilitation. By portraying the energy state during and post-crisis clearly, energy models enable policymakers and utilities to make intelligent decisions that enhance rapid and efficient restoration and long-term energy security. Table 6.10 makes reference to the various uses of energy system models in averting different forms of energy crises ranging from natural to geopolitical crises.

6.7.3. *Building robust energy systems for future generations*

In the future, developing strong energy systems is crucial to maintaining long-term energy security and sustainability. Energy system models help in developing systems which will be capable to counter new challenges, such as rising technology, shifting energy needs and climatic effects (Arun et al. 2025). Models explore how existing infrastructure can be made more robust to handle disruptions, such as integrating hydrogen fuel cells, next-generation nuclear reactors and smart grids to increase flexibility. They also look at energy storage alternatives, including large batteries, to balance supply and demand during periods when renewable energy is low. Looking forward through trends, including increases in EV use or declining fossil fuel prices, these models guide investment in technologies and infrastructure that commit to energy systems now and in the future. Lastly, they call for the creation of flexible, sustainable and fair energy systems that supply affordable, clean and secure energy to future generations.

6.8. Conclusion

6.8.1. *Summary of key findings*

This chapter emphasizes the key role of energy system models in shaping the future of world energy. They forecast energy transitions, enable the integration of renewable energy, reduce greenhouse gas emissions and improve energy security.

By extrapolating various scenarios and examining the impacts of policies, technologies and economic trends, they provide useful information for decision-makers to plan for a sustainable energy future. Comprehensive strategies, that is, bottom-up, top-down and hybrid models, are required to address the entire range of interactions between energy systems, the economy, society and the environment. Energy system models play an important role in identifying cost-optimal pathways to climate goals such as net-zero emissions by 2050 and are useful to offer insights for long-term investment. They also prepare for shock, such as natural or geopolitical shock, by making energy resilience possible through strategies that enable energy transitions and reductions in emissions. Danish, British and Californian case studies show how the models have made energy transitions and emissions reductions successful.

6.8.2. *The future of energy system modeling*

Aspect	Future contributions
Energy security	Real-time monitoring and adaptive modeling to anticipate supply disruptions Scenario analysis for diverse and resilient energy sources Integration of geopolitical risks and resource dependencies
Decentralization	Modeling distributed energy resources like rooftop solar, microgrids and storage Assessing impacts of peer-to-peer energy trading and local markets Supporting planning for flexible, decentralized grids
Sustainability	Detailed emissions and environmental impact forecasting Evaluating clean technology adoption pathways Integrating climate policy targets with energy planning
Efficiency and innovation	Leveraging AI and machine learning for dynamic forecasting and optimization Faster computation enabling scenario analysis of complex, multi-sector systems Modeling emerging technologies and their system integration
Social inclusion and equity	Incorporating behavioral and socioeconomic variables to ensure just transitions Identifying vulnerable populations affected by energy changes Supporting policies for equitable access and benefits
Resilience and flexibility	Simulating system responses to extreme weather and crises Designing adaptable energy systems that respond to changing conditions Enhancing multi-sector integration (electricity, heat, transport) for robustness

Table 6.11. *Role of future energy system models in addressing energy challenges*

In the future, energy system modeling will provide greater sustainability, resilience and efficiency in global energy. More advanced models will be needed for more complex energy systems with more renewables, decentralized power generation and new technologies. Better modeling approaches and computing capabilities will allow for better forecasts and more intelligent decisions.

As energy security demand and climate concerns grow, such models will be important to inform energy-emissions-reduction policy, efficiency improvement and energy access for all. Big data, AI and machine learning technologies will make models dynamic and responsive in real-time. Pressure is also mounting to include social, economic and behavioral variables in models so that energy transitions are inclusive and equitable.

The next generation of models must address technology priorities as well as the social impacts of energy policy. It is through this integral consideration that holistic design for future sustainable and resilient energy systems can be developed.

Table 6.11 gives an overview of how future energy system models will support various key areas of the energy transition. It emphasizes their function in improving energy security, facilitating decentralization, enhancing sustainability and providing social equity.

6.8.3. *Policy recommendations and next steps*

Energy system models can offer useful guidance to develop sustainable and resilient energy systems. Governments and energy decision-makers need to first invest in further developing these models to support long-term planning and smart decision-making. The models should be open and shareable so that they are increasingly used by more people in conjunction with one another. Second, the policymakers need to use these tools to create policies that cover the whole energy system, aiming at net-zero emissions while keeping the energy affordable and the economy secure. Third, models can find out the principal technologies like carbon capture and energy storage, hence it is imperative to invest in innovation in those sectors. Fourth, transitions need to be fair transitions experimenting with how policies affect different groups and make sure everyone has access to clean, affordable energy. And finally, energy models can allow countries to cooperate through data sharing and solutions in fighting climate change and energy poverty. As models become more advanced, they will become even more essential in creating strong, sustainable energy systems for the future.

6.9. References

Ahmadizadeh, M., Heidari, M., Thangavel, S., Naamani, E.A., Khashehchi, M., Verma, V., Kumar, A. (2024). Technological advancements in sustainable and renewable solar energy systems. In *Highly Efficient Thermal Renewable Energy Systems: Design, Optimization and Applications*, Verma, V., Thangavel, S., Dutt, N., Kumar, A., Weerasinghe, R. (eds). CRC Press, Boca Raton, 23–39. doi: 10.1201/9781003472629-2.

Alamoush, A.S., Ballini, F., Ölçer, A.I. (2024). Management of stakeholders engaged in port energy transition. *Energy Policy*, 188, 114074. doi: 10.1016/j.enpol.2024.114074.

Anbalagan, N., Akilandeswari, J., Mouliraj, V., Nandhini, P., Kalaivani, P., Perumal, I. (2024). Envisioning India's energy future: Predictive models for power generation trends. In *2024 International Conference on Emerging Smart Computing and Informatics (ESCI)*. IEEE, Pune. doi: 10.1109/esci59607.2024.10497272.

Arun, M., Samal, S., Barik, D., Chandran, S.S., Tudu, K., Praveenkumar, S. (2025). Integration of energy storage systems and grid modernization for reliable urban power management toward future energy sustainability. *Journal of Energy Storage*, 114, 115830. doi: 10.1016/j.est.2025.115830.

Ashraf, A. and Sagheer, M. (2025). Renewable energy capacity and technological innovations: A review of global trends and future directions. *Environmental Progress & Sustainable Energy*, 44(6), e70071. doi: 10.1002/ep.70071.

Bakhsh, S., Zhang, W., Ali, K., Oláh, J. (2024). Strategy towards sustainable energy transition: The effect of environmental governance, economic complexity and geopolitics. *Energy Strategy Reviews*, 52, 101330. doi: 10.1016/j.esr.2024.101330.

Behrens, J., Zeyen, E., Hoffmann, M., Stolten, D., Weinand, J.M. (2024). Reviewing the complexity of endogenous technological learning for energy system modeling. *Advances in Applied Energy*, 100192. doi: 10.1016/j.adapen.2024.100192.

Cedeno, R.C.B. and Wei, J. (2024). Investigating factors to promote energy: Socio-political barriers to renewable energy sector in India. *The Electricity Journal*, 37(2), 107366. doi: 10.1016/j.tej.2024.107366.

Cieplinski, A., D'Alessandro, S., Marghella, F. (2021). Assessing the renewable energy policy paradox: A scenario analysis for the Italian electricity market. *Renewable and Sustainable Energy Reviews*, 142, 110838. doi: 10.1016/j.rser.2021.110838.

Craig, J. and Latham, A. (2024). Carbon matters: Fossil fuels and the energy transition. *EPJ Web of Conferences*, 310, 00002. doi: 10.1051/epjconf/202431000002.

Dabla, N.E. and Goldthau, A.C. (2023). Clean energy services: Universal access as enabler for development? In *The Environment in Global Sustainability Governance*, Partzsch, L. (ed.). Bristol University Press, Bristol, 174–193. doi: 10.56687/9781529228021-017.

Danish, M.S.S. and Senjyu, T. (2023). Shaping the future of sustainable energy through AI-enabled circular economy policies. *Circular Economy*, 2(2), 100040. doi: 10.1016/j.cec.2023.100040.

Das, D., Chakraborty, I., Bohre, A.K., Kumar, P., Agarwala, R. (2024). Sustainable integration of green hydrogen in renewable energy systems for residential and EV applications. *International Journal of Energy Research*, 1–20. doi: 10.1155/2024/8258624.

Denysov, V., Babak, V., Zaporozhets, A., Nechaieva, T., Kostenko, G. (2024). Energy system optimization potential with consideration of technological limitations. In *Nexus of Sustainability*, Zagorodny, A., Bogdanov, V., Zaporozhets, A. (eds). Springer, Geneva, 113–126. doi: 10.1007/978-3-031-66764-0_5.

Dharmapriya, N., Edirisinghe, S., Gunawardena, V., Methmini, D., Jayathilaka, R., Dharmasena, T., Wickramaarachchi, C., Rathnayake, N. (2024). Towards a greener future: Examining carbon emission dynamics in Asia amid gross domestic product, energy consumption, and trade openness. *Environmental Science and Pollution Research*, 31(14), 21488–21508. doi: 10.1007/s11356-024-32475-y.

Ekechukwu, N.D.E. and Simpa, N.P. (2024). The future of cybersecurity in renewable energy systems: A review, identifying challenges and proposing strategic solutions. *Computer Science & IT Research Journal*, 5(6), 1265–1299. doi: 10.51594/csitrj.v5i6.1197.

Emrani, A. and Berrada, A. (2024). Modeling and optimal capacity configuration of dry gravity energy storage integrated in off-grid hybrid PV/Wind/Biogas plant incorporating renewable power generation forecast. *Journal of Energy Storage*, 97, 112698. doi: 10.1016/j.est.2024.112698.

Evro, S., Oni, B.A., Tomomewo, O.S. (2024). Carbon neutrality and hydrogen energy systems. *International Journal of Hydrogen Energy*, 78, 1449–1467. doi: 10.1016/j.ijhydene.2024.06.407.

Fodstad, M., Del Granado, P.C., Hellemo, L., Knudsen, B.R., Pisciella, P., Silvast, A., Bordin, C., Schmidt, S., Straus, J. (2022). Next frontiers in energy system modelling: A review on challenges and the state of the art. *Renewable and Sustainable Energy Reviews*, 160, 112246. doi: 10.1016/j.rser.2022.112246.

Fotopoulou, M., Rakopoulos, D., Petridis, S., Drosatos, P. (2023). Assessment of smart grid operation under emergency situations. *Energy*, 287, 129661. doi: 10.1016/j.energy.2023.129661.

Gupta, L., Sarkar, B.D., Shankar, R. (2025). Strategic modelling for sustainable power grids: A comprehensive study on energy storage integration in India. *Journal of Modelling in Management*. doi: 10.1108/jm2-02-2024-0060.

Hassan, Q., Viktor, P., Al-Musawi, T.J., Ali, B.M., Algburi, S., Alzoubi, H.M., Al-Jiboory, A.K., Sameen, A.Z., Salman, H.M., Jaszczur, M. (2024). The renewable energy role in the global energy Transformations. *Renewable Energy Focus*, 48, 100545. doi: 10.1016/j.ref.2024.100545.

Heras, A. and Gupta, J. (2023). Fossil fuels, stranded assets, and the energy transition in the Global South: A systematic literature review. *Wiley Interdisciplinary Reviews Climate Change*, 15(1). doi: 10.1002/wcc.866.

Hu, H., Yu, S.S., Trinh, H. (2024). A review of uncertainties in power systems modeling, impact, and mitigation. *Designs*, 8(1), 10. doi: 10.3390/designs8010010.

Joshi, P., Rao, A.B., Banerjee, R. (2025). Review of solar PV deployment trends, policy instruments, and growth projections in China, the United States, and India. *Renewable and Sustainable Energy Reviews*, 213, 115436. doi: 10.1016/j.rser.2025.115436.

Kaplan, P.O., Boyd, G., Browning, M., Perl, K., Supekar, S., Victor, N., Worrell, E. (2025). Is the industrial sector hard to decarbonize or hard to model? A comparative analysis of industrial modeling and net zero carbon dioxide pathways. *Energy and Climate Change*, 100190. doi: 10.1016/j.egycc.2025.100190.

Keirstead, J., Jennings, M., Sivakumar, A. (2012). A review of urban energy system models: Approaches, challenges and opportunities. *Renewable and Sustainable Energy Reviews*, 16(6), 3847–3866. doi: 10.1016/j.rser.2012.02.047.

Krishna, B.S., Pauline, S., Sivakumar, S., Palagan, C.A., Talasila, V., Krishna, M.M.S. (2024). Enhanced efficiency in smart grid energy systems through advanced AI-based thermal modeling. *Thermal Science and Engineering Progress*, 53, 102765. doi: 10.1016/j.tsep.2024.102765.

Modis, T. (2022). Forecasting the mix of world energy needs by mid-twenty-first century. In *Global Challenges of Climate Change, Vol. 1*, Devezas, T.C., Leitão, J.C.C., Yegorov, Y., Chistilin, D. (eds). Springer, Cham, 125–137. doi: 10.1007/978-3-031-16470-5_9.

Montant, G. (2025). The effectiveness of OPEC and OPEC+ from 2009 to 2024: An empirical appraisal. *Resources Policy*, 103, 105529. doi: 10.1016/j.resourpol.2025.105529.

Newell, R. and Iler, S. (2013). *The Global Energy Outlook*. National Bureau of Economic Research, Cambridge, MA. doi: 10.3386/w18967.

Newell, R., Qian, Y., Raimi, D. (2016). *Global Energy Outlook 2015*. National Bureau of Economic Research, Cambridge, MA. doi: 10.3386/w22075.

Nicoli, M., Faria, V.A.D., De Queiroz, A.R., Savoldi, L. (2024). Modeling energy storage in long-term capacity expansion energy planning: An analysis of the Italian system. *Journal of Energy Storage*, 101, 113814. doi: 10.1016/j.est.2024.113814.

Pfenninger, S., Hawkes, A., Keirstead, J. (2014). Energy systems modeling for twenty-first century energy challenges. *Renewable and Sustainable Energy Reviews*, 33, 74–86. doi: 10.1016/j.rser.2014.02.003.

Ross-Hopley, D., Rahman, S., Ugwu, L., Ibrahim, H. (2025). Techno-economic opportunities for integration of renewable energy into the Saskatchewan energy system using EnergyPLAN. *Energy*, 134903. doi: 10.1016/j.energy.2025.134903.

Shoaei, M., Noorollahi, Y., Hajinezhad, A., Moosavian, S.F. (2024). A review of the applications of artificial intelligence in renewable energy systems: An approach-based study. *Energy Conversion and Management*, 306, 118207. doi: 10.1016/j.enconman.2024.118207.

Sridhar, S. and Salkuti, S.R. (2022). Development and future scope of renewable energy and energy storage systems. *Smart Cities*, 5(2), 668–699. doi: 10.3390/smartcities5020035.

Veers, P., Dykes, K., Basu, S., Bianchini, A., Clifton, A., Green, P., Holttinen, H., Kitzing, L., Kosovic, B., Lundquist, J.K. et al. (2022). Grand challenges: Wind energy research needs for a global energy transition. *Wind Energy Science*, 7(6), 2491–2496. doi: 10.5194/wes-7-2491-2022.

Wu, L. and Hussain, S. (2025). Navigating the energy transition: Interplay of geopolitics, economic complexity, and environmental governance in OECD countries. *Energy Strategy Reviews*, 57, 101624. doi: 10.1016/j.esr.2024.101624.

Zaiter, I., Ramadan, M., Bouabid, A., Mayyas, A., El-Fadel, M., Mezher, T. (2024). Enabling industrial decarbonization: Framework for hydrogen integration in the industrial energy systems. *Renewable and Sustainable Energy Reviews*, 203, 114782. doi: 10.1016/j.rser.2024.114782.

7

Application of Digital Technologies in Renewable Energy Systems

The transition to renewable energy system (RES) requires digital technologies to improve efficiency, reliability and integration. Digital approaches such as artificial intelligence (AI), machine learning (ML), the Internet of Things (IoT), blockchain, digital twins, energy management system (EMS), cybersecurity and cloud computing are revolutionizing and transforming energy system. AI and ML improve forecasting accuracy by up to 20%, enable early fault detection, optimize energy distribution and predictive maintenance in both solar and wind systems. IoT devices support real-time monitoring, automated control and grid stability, contributing to a 2–5% increase in overall system efficiency. Blockchain facilitates secure, transparent energy transaction and enables peer-to-peer (P2P) energy trading, decentralized energy market and microgrid management with up to 15% reduction in transaction and settlement time.

Digital twins analyze performance optimization, fault prediction and diagnostics, system planning and expansion, and create virtual models of energy system. EMS provides optimal energy generation, storage and consumption for home and industries, grid and demand side management using software-based automation, resulting in 10–30% improvements in energy utilization and demand-side efficiency. Cybersecurity protects renewable energy infrastructure from cyberattacks and unauthorized access to secure energy trading, grid safety and IoT security. By storing critical insights in cloud platform, cloud computing processes energy data locally for faster decision-making and remote monitoring. Integration of AI, IoT, blockchain and cloud computing is paving the way for smart, decentralized and secure energy networks. As the global energy landscape shifts toward sustainability, these digital innovations address challenges such as grid stability, energy variability and system inefficiency, which promotes more sustainable and reliable RES. In sections 7.3, 7.4, 7.5 and 7.7 of this chapter, the analysis and implementation of several digital approaches in renewable power systems that uses solar and wind energy are explained in detail.

Chapter written by Manimegalai S. and Binu Ben Jose D.R.

7.1. Introduction

Digitalization has emerged as a fundamental catalyst in the modern energy transition, reshaping the development, operation and integration of RESs into the broader energy landscape. As countries transition from traditional fossil fuel-based infrastructures to low-carbon renewable energy driven systems, they face significant operational challenges due to the complexity, variability and decentralization inherent in renewable sources such as solar, wind and biomass. Managing these dynamic systems requires not only robust physical infrastructure but also advanced digital solutions. Digital technologies including AI, IoT, cloud computing and advanced data analytics are increasingly essential for providing real time visibility, intelligent control and seamless grid integration. From solar photovoltaic (PV) arrays to offshore wind farms and distributed energy storage, digitalization enables continuous performance monitoring, generation and demand forecasting, maintenance prediction and optimization of energy flow across decentralized networks. This approach enhances system efficiency, reliability and supports the transition toward a sustainable and resilient energy future.

7.1.1. *Key contributions of digitalization*

7.1.1.1. *Real-time monitoring and control*

Digital sensors, IoT networks and supervisory control and data acquisition (SCADA) systems provide real-time data on energy generation, consumption and asset performance. These data are used to balance supply and demand, detect anomalies and improve response times in decentralized systems (Bentarzi et al. 2018). SCADA systems enable centralized monitoring of RES. IoT sensors provide continuous feedback on voltage, temperature, current and output. Edge computing enhances localized decision-making with minimal latency.

7.1.1.2. *Renewable integration and grid flexibility*

Renewable sources such as solar and wind are intermittent. AI and digital twins help predict generation patterns and adjust grid operations dynamically. These tools allow for flexible energy dispatch, storage management and demand response strategies (International Renewable Energy Agency (IRENA) 2022). It enables seamless coordination between distributed generators, storage systems and consumers. Two-way communication allows energy flow adjustments in real-time. Demand-side management (DSM) programs use digital control for load shifting and peak saving. AI and ML algorithms improve the accuracy of solar irradiance and wind speed predictions. Satellite data and remote sensing are used for site selection and yield estimation. Real-time weather integration into energy dispatch models enables dynamic planning.

7.1.1.3. *Energy storage, battery management and simulation*

AI-based battery management systems (BMS) optimize charge/discharge cycles. State of health (SOH) and state of charge (SOC) are predicted using data-driven models. Integration with solar and wind systems smoothens intermittent generation. Energy modeling software (e.g. HOMER, MATLAB-based apps) supports system sizing and techno-economic analysis. Geographic information system (GIS)-based tools are used for spatial planning of solar, wind and hybrid energy installations, while digital scenario simulations support comprehensive policy design and informed investment planning (Vagnoni et al. 2021).

7.1.1.4. *Predictive maintenance*

Digitalization facilitates predictive maintenance through ML algorithms that analyse sensor data to forecast component failures. This reduces downtime, cuts maintenance costs and extends asset life, especially in remote or large-scale solar and wind farms (Onwusinkwue et al. 2024). ML models detect early signs of wear and faults in PV panels, wind turbines and batteries. Digital twins simulate system behavior and provide maintenance insights that help optimize repair scheduling, reduce operational costs and maximize asset uptime.

7.1.1.5. *Decentralized energy markets and prosumers*

Blockchain and smart contracts are enabling decentralized energy trading. Households with rooftop solar and batteries can sell surplus energy directly to neighbors, promoting local energy resilience and economic participation (Mengelkamp et al. 2018). Blockchain enables P2P energy exchanges, particularly in microgrids. Smart contracts automate billing, payments and verification, enabling trusted, secure and autonomous transactions. This digital framework empowers consumers to become active participants in the energy ecosystem by generating, consuming and trading electricity as prosumers.

7.1.1.6. *Decision support and policy*

Digital tools such as GIS, energy simulation models and AI-based policy analysis are being used to plan renewable deployments, assess impacts and guide energy policy with evidence-based insights (Rehman et al. 2023). Mobile apps and dashboards allow consumers to monitor usage, generation and savings. Engagement-driven digital platforms and real-time feedback tools encourage energy conservation behavior and informed decision-making. Utilities use digital insights for customer segmentation and the design of targeted energy efficiency programs.

7.1.2. *Objective*

Renewable energy	Digital technologies	Applications	Benefits	Improvement in literature (%)
Solar energy system	AI/ML, IoT sensors, smart meters, SCADA, digital twins	Soiling and shading detection, inverter fault diagnosis, smart energy monitoring	Increased energy yield, lower maintenance cost	15–25% increase in yield due to predictive and fault detection
Wind energy systems	Vibration sensors, edge AI, ML-based, fault detection, digital twins	Gearbox and blade health monitoring, predictive maintenance, yaw control	Reduced downtime, longer component life, enhanced energy capture	20–30% reduction in unplanned outages
Hydro-power systems	SCADA, AI forecasting models, remote sensing, sensor networks	Flow rate and reservoir level prediction, real-time turbine control, dam safety alerts	Improved safety and reliability, optimized water usage	10–20% improvement in operational efficiency and safety
Biomass/biogas plants	IoT, AI/ML, combustion modeling software, gas composition analyzers	Feedstock quality monitoring, combustion efficiency control, emission tracking	Consistent power output, reduced fuel waste, compliance with emission standards	15–25% fuel efficiency gain and emission reduction
Hybrid Systems	EMS, blockchain, digital twins forecasting algorithms	Real-time energy source switching, storage and grid coordination, load balancing	Increased system reliability, reduced fossil backup, lower operational cost	20–35% increase in system efficiency and cost savings

Table 7.1. *Integration of digital technologies across renewable energy systems*

This chapter aims to provide readers with a comprehensive understanding of how digitalization is transforming the renewable energy landscape. It introduces key digital technologies such as AI, IoT, blockchain, SCADA systems and digital twins to examine their specific applications across renewable energy domains including solar PV, wind, hydropower, biomass, and hybrid systems. The chapter will explore how AI-driven predictive maintenance and real-time performance monitoring contribute to enhance asset reliability and operational efficiency. Real-world case studies will be analyzed to demonstrate how these technologies are being practically deployed to support the global energy transition. Additionally, it will address critical challenges such as cybersecurity threats, lack of interoperability and skill shortages that hinder widespread adoption. Finally, the chapter presents future directions including autonomous systems and P2P energy markets and offers strategic recommendations for policymakers, industry stakeholders and researchers to scale up digital innovations in renewable energy. The integration of these technologies enables intelligent forecasting, system optimization and enhanced control, as summarized in Table 7.1.

7.2. Overview of RESs

RESs from naturally replenished sources such as solar radiation, wind and biomass offer sustainable alternatives to fossil fuels. RES technologies have been seen in common deployment with the growing global emphasis on clean energy such as PV, wind energy plus biomass-based systems. As of 2024, global capacities have risen to approximately 1,400 GW in terms of solar PV, to 1,020 GW in terms of wind, and to 150 GW in terms of biomass and India also contributes through 81.8 GW for solar, 45.8 GW for wind, plus 10.2 GW for biomass installations as shown in Table 7.2.

Energy systems integrating solar and wind with storage are gaining traction due to their improved reliability and efficiency. According to comparative assessments, solar and wind are intermittent and have moderate capacity factors (18–35%), but biomass offers high dispatchability as well as capacity (70–85%), so it is indeed suitable for continuous power generation. Hybrid systems balance intermittency. Understanding these variations is required for optimized design and policy planning. Transitioning to a low-carbon energy future depends upon it. A comparative summary of these performance parameters is provided in Table 7.3, which highlights the technical advantages and limitations of each RES type.

Technology	Global installed capacity (GW)	India installed capacity (GW)
Solar PV	1,400	81.8
Wind energy	1,020	45.8
Biomass power	150	10.2
Hybrid systems	~120 (est.)	10.2

Table 7.2. *Installed capacity statistics: global and India (renewable energy)*

7.2.1. *Solar PV systems*

Solar PV systems convert sunlight into electricity using semiconductor-based PV cells. They use semiconductor-based PV cells, typically made from silicon that absorb photons from sunlight and release electrons creating an electric current. This current is initially in the form of direct current (DC) and must be converted into alternating current (AC) using an inverter before it can be used for homes, businesses or fed into the electricity grid as shown in Figure 7.1. The solar PV systems are highly scalable and are found in applications ranging from small rooftop installations to massive utility-scale solar farms. These systems are widely deployed due to their modularity, declining costs and minimal operational requirements. Rooftop PV enables distributed generation, while utility-scale solar plants feed power directly into national grids. India's *Rewa Ultra Mega Solar Project* in Madhya Pradesh is a leading example of a utility-scale solar farm, generating 750 MW and supplying clean energy to Delhi Metro and state consumers (World Bank 2020).

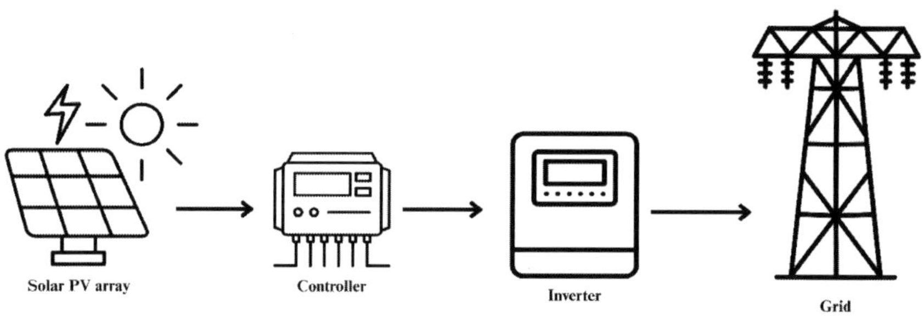

Figure 7.1. *Overview of solar PV system*

7.2.2. *Wind energy systems*

Wind energy systems produce electricity by converting the kinetic energy of moving air into mechanical energy and subsequently into electrical energy. Wind turns the blades of a turbine causing a rotor to spin. The rotor is connected to a main shaft, which turns a generator. Inside the generator an electromagnetic induction produces AC electricity as shown in Figure 7.2. Wind turbines are installed in both onshore and offshore locations. Onshore systems are more widely adopted due to easier accessibility, lower installation costs and simpler maintenance. Although offshore wind farms are more complex and costly to develop, they benefit from stronger and more consistent wind patterns, resulting in higher and more stable energy output. Advanced technologies such as AI and digital twins are increasingly used in both settings to enhance performance monitoring, enable early fault detection and support predictive maintenance strategies. For example, The *Hornsea One* project in the UK is the world's largest offshore wind farm, with a capacity of 1.2 GW. It uses digital twin technology and predictive analytics for turbine health and efficient maintenance (Ørsted 2021).

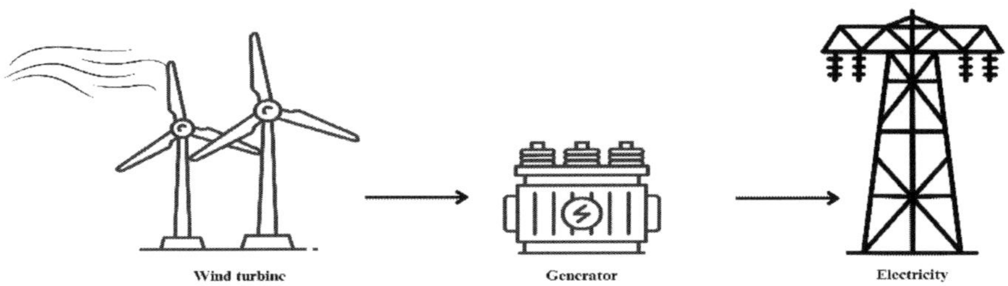

Figure 7.2. *Overview of a wind energy system*

7.2.3. *Hydro-electric power systems*

Hydro-electric power systems harness the energy of flowing or falling water to generate electricity. In these systems, water is channelled through dams, penstocks or natural river flows to spin the blades of a turbine. The turbine's mechanical energy drives a generator, which produces electrical energy through electromagnetic induction as shown in Figure 7.3. Small-scale hydropower plants serve localized communities, while large-scale hydroelectric dams can supply significant portions of national power grids. Hydropower is valued for its ability to provide steady, base-load power and rapidly respond to fluctuations in energy demand. Large dams offer reliable base-load power, while small hydropower systems serve decentralized or rural needs. For example, Norway's *Alta Hydropower Plant* integrates real-time weather and inflow data to optimize water usage and reduce flood risks, supported by AI-based control systems (Arch et al. 2019).

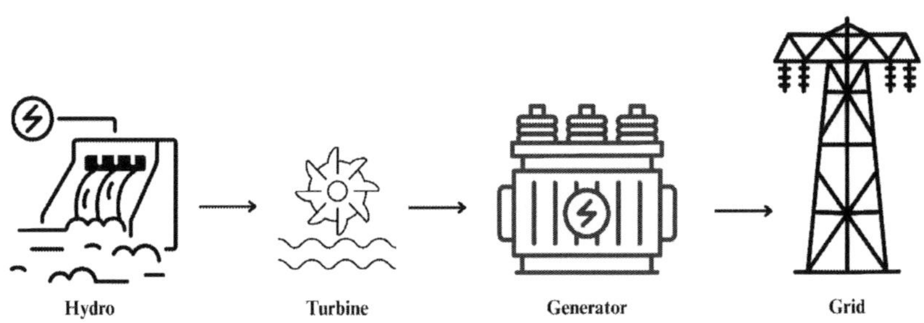

Figure 7.3. *Overview of a hydro-electric power system*

7.2.4. *Biomass and biogas systems*

Biomass energy comes from organic materials such as wood, crop waste and animal dung, while biogas is produced through anaerobic digestion. Biomass and biogas systems convert organic materials into electricity using different processes. Biomass such as wood chips, agricultural residues or organic waste is combusted to produce heat. This heat generates steam that drives a steam turbine connected to an electrical generator. In biogas systems the organic material undergoes anaerobic digestion, a biological process that produces methane-rich gas. This biogas is then combusted in a gas engine or turbine to generate electricity, as shown in Figure 7.4. Both systems offer a way to produce renewable energy while simultaneously managing waste and reducing greenhouse gas emissions. These systems provide a renewable solution for both heat and power generation, especially in agricultural and rural areas. Smart sensors are used to optimize combustion efficiency and continuously monitor emissions for compliance and environmental safety. For example, Germany's *Bioenergie Dorf Jühnde* is a pioneering biomass village where biogas is generated from manure and used for electricity and district heating. IoT systems track gas composition and system efficiency (IEA Bioenergy 2021).

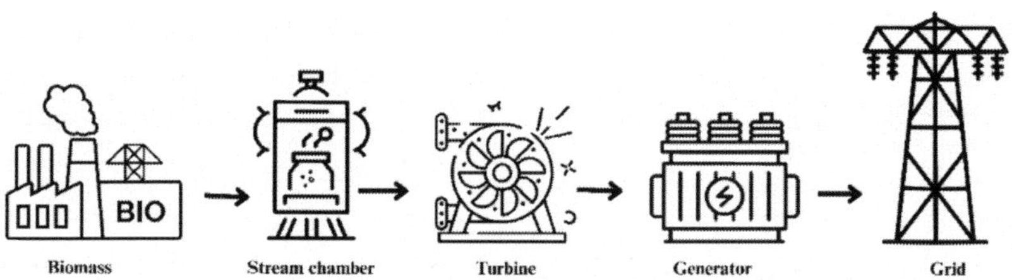

Figure 7.4. *Scheme of a biomass electric energy conversion system*

7.2.5. *Hybrid and microgrid systems*

Hybrid systems combine different energy sources (e.g. solar + battery + diesel) to ensure reliability, especially in areas without stable grid connections. Hybrid and microgrid systems combine multiple energy sources such as solar, wind, diesel generators and battery storage to ensure a continuous and reliable supply of electricity, as shown in Figure 7.5. Solar panels and wind turbines generate electricity as described above, while batteries store excess energy for use during periods of low generation. An EMS coordinates the interaction between these sources, optimizing performance and ensuring seamless power delivery. Microgrids can operate independently or in conjunction with the main utility grid making them essential for rural electrification, disaster recovery and enhancing grid resilience. Microgrids allow for local energy independence and are digitally controlled using EMSs and remote SCADA monitoring. The *Smart Power India microgrid initiative*, supported by the Rockefeller Foundation, has deployed solar-battery-diesel hybrid microgrids in over 250 villages in Uttar Pradesh and Bihar, bringing reliable power through remote monitoring and AI-based load forecasting (Rockefeller Foundation 2022).

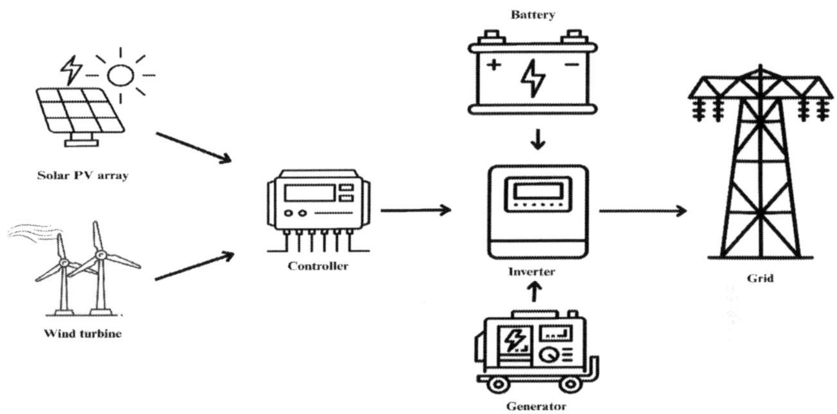

Figure 7.5. *Electricity generation through a hybrid system*

Parameter	Solar PV	Wind energy	Biomass-energy	Hybrid
Capacity factor (%)	18–22	25–35	70–85	40–60
CO_2 emissions (g/kWh)	~45 (incl. lifecycle)	~11	35 – 90	<20 (optimized systems)
Land requirement (MW/km^2)	35–60	2–3	5–10	Variable (lower when integrated)
Dispatchability	Intermittent	Seasonal/intermittent	Dispatchable	High with storage
Storage requirement	High (for autonomy)	Medium–high	Low	Optimized
O&M cost (INR/kWh)	0.15–0.25	0.2–0.3	0.4–0.6	0.3–0.5

Table 7.3. *Comparative performance of RES technologies*

7.3. Core digital technologies

The transition to intelligent and autonomous RES is being powered by a suite of core digital technologies. These technologies form the foundation of Industry 4.0 and enable real-time decision-making, predictive analytics, automation and enhanced operational efficiency. The most important digital technologies relevant to modern energy systems are shown in Figure 7.6.

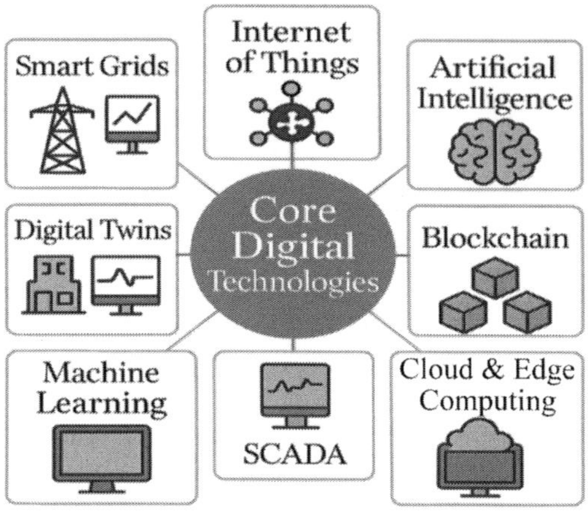

Figure 7.6. *Core digital technologies*

7.3.1. *Hybrid and microgrid systems: enabling two-way energy and data flow*

A smart grid is an advanced electricity network that uses digital communication and automation technologies to manage the flow of electricity and data between energy producers and consumers. Unlike traditional grids, which only allow one-way power delivery from the utility provider to the consumer, smart grids enable two-way flow of both electricity and information. This allows real-time monitoring, quick fault detection, automatic rerouting of power during outages and better integration of variable renewable energy sources like solar and wind as shown in Table 7.4. They support demand response, where energy usage can be adjusted based on availability and pricing by enhancing overall grid flexibility and reliability.

Key functions are as follows:

– integrates distributed energy sources (solar panels and wind turbines);

– balances supply and demand dynamically;

– enables real-time pricing and load control;

– detects outages and restores service faster.

Sr. No.	Component/approach	Application/system considered	Reference
1.	Advanced metering infrastructure (AMI)	Real-time load monitoring and dynamic pricing	Udo et al. 2023
2.	Phasor measurement units (PMUs)	Grid stability and voltage profile management	Bentarzi et al. 2018
3.	Demand response mechanisms	Peak load management in smart microgrids	International Energy Agency (2023)
4.	Wide area monitoring systems (WAMS)	Renewable integration and fault detection	Liao et al. 2022
5.	IEC 61850 Protocol	Interoperability in smart substations	Hameed et al. 2020

Table 7.4. *Smart grid technologies*

7.3.2. *IoT and sensor networks: real-time monitoring of renewable assets*

The IoT in energy refers to the use of connected devices and sensors that collect and transmit data in real-time. In renewable systems, the IoT devices are installed on solar panels, wind turbines, batteries and inverters to monitor performance parameters such as temperature, output voltage and equipment health, as presented in Table 7.5. These devices feed data into central platforms or edge computing systems, which enable operators to optimize performance, schedule maintenance and prevent failures before they occur.

Key functions are as follows:

– remote monitoring and control of assets;

– data-driven performance optimization;

– early fault detection and diagnostics;

– integration with SCADA and cloud systems.

Sr. No.	Component/approach	Application/system considered	Reference
1.	Wireless sensor networks (WSNs)	Soil moisture and sunlight tracking in solar farms	Lu et al. 2019
2.	Low power wide area network (LoRa)	Remote RES performance monitoring	Al-Shetwi et al. 2025
3.	Message Queuing Telemetry Transport (MQTT) Protocol	Lightweight IoT communication in microgrids	Bentarzi et al. 2018
4.	Smart meters	Two-way power and usage feedback	International Energy Agency (2023)
5.	Edge IoT gateways	Real-time data preprocessing in off-grid systems	Qing and Niu 2018

Table 7.5. *IoT and sensor networks*

7.3.3. *Artificial intelligence and machine learning: forecasting, optimization and predictive analytics*

AI and ML have emerged as powerful tools in the digital transformation of RES. These technologies are capable of processing and analysing vast datasets collected from IoT sensors, weather forecasting systems and historical energy usage records. Their applications are wide-ranging and critical to system optimization. In particular, AI enhances the accuracy of forecasting solar irradiance, wind speed and electricity demand, which are essential for balancing supply and demand in variable renewable energy environments, as presented in Table 7.6. It also supports intelligent energy dispatch in hybrid systems by evaluating multiple inputs in real-time. Furthermore, AI-driven predictive maintenance enables the early detection of anomalies and equipment faults by reducing unplanned downtime and extending asset life. Additionally, AI and ML provide decision support for microgrid control, energy trading and dynamic participation in energy markets, thereby contributing to greater operational efficiency, economic value and system resilience.

Key functions are as follows:

– improves accuracy in generation prediction;

– automates energy scheduling and control;

– minimizes downtime through intelligent fault prediction.

Sr. No.	Technique/algorithm	Application/system considered	Reference
1.	Artificial neural networks (ANN)	Solar irradiance and load forecasting	Qing and Niu 2018
2.	Support vector machines (SVM)	Fault detection in wind turbines	Kusiak and Verma 2011
3.	Reinforcement learning	Energy dispatch optimization in microgrids	Zang et al. 2024
4.	K-Means clustering	Load pattern segmentation	Jahangir et al. 2020
5.	Decision trees/random forest	Predictive maintenance in hybrid RES	Udo et al. 2023

Table 7.6. *Artificial intelligence and machine learning*

7.3.4. *Digital twins: simulation and virtual modeling of physical systems*

A digital twin is a real-time virtual replica of a physical renewable energy asset (e.g. wind turbine and solar farm). It simulates the behavior performance and health of the actual asset under various conditions using real-time data. Digital twins are used for testing different operational scenarios, predicting failures and planning maintenance without affecting the real system, as presented in Table 7.7.

Key functions are as follows:

– visual simulation of energy systems;

– scenario testing and risk analysis;

– enhanced asset management and lifecycle planning.

Sr. No.	Component/feature	Application/system considered	Reference
1.	Real-time simulation models	Predictive monitoring of wind farms	Foley et al. 2012
2.	Virtual sensor integration	Cost-effective fault prediction in PV systems	Deitsch et al. 2019
3.	System optimization twin	Grid efficiency improvement	Liao et al. 2022
4.	Asset lifecycle management	O&M scheduling in distributed RES	Banik and Biswas 2024
5.	Twin with AI integration	Self-learning adaptive control	Al-Shetwi et al. 2025

Table 7.7. *Digital twins*

7.3.5. *Blockchain: peer-to-peer energy trading and smart contracts*

Blockchain technology provides a secure and transparent way to record energy transactions. In RES, it enables P2P energy trading by allowing prosumers (producer-consumers) to sell excess solar or wind power to neighbors directly. Smart contracts automate these transactions and ensure trust without a central authority as presented in Table 7.8. Blockchain also supports green energy certification and ensures traceability of renewable sources.

Key functions are as follows:

– decentralized and transparent energy trading;

– tamper-proof energy transaction records;

– supports local energy markets and microgrids.

Sr. No.	Blockchain feature/tool	Application/system considered	Reference
1.	Smart contracts	Peer-to-peer energy trading	Zhang et al. 2024
2.	Distributed ledger technology	Secure transaction logging in VPPs	Lu et al. 2019
3.	Hyperledger fabric framework	Energy asset registration	Lu et al. 2019
4.	Tokenization	Incentivizing prosumers in smart grids	Lee and Choi 2019
5.	Blockchain-IoT integration	Secure sensor data for solar farms	Lu et al. 2019; Zhang et al. 2024

Table 7.8. *Blockchain*

7.3.6. *SCADA and remote monitoring systems: supervisory control and automation*

SCADA systems are integral to the operation and management of renewable energy facilities, including solar farms, wind parks, hydropower stations and microgrids. These systems collect real-time data from sensors, meters and control devices deployed across the energy network, allowing centralized monitoring and control through a user-friendly interface as presented in Table 7.9. SCADA enables operators to observe system health, track performance metrics and respond to alarms or faults promptly. It supports both manual and automated control actions such as adjusting turbine settings or switching inverters based on live operational data. In addition to real-time oversight, SCADA systems provide historical data logging which facilitates trend analysis, predictive maintenance and performance optimization. When integrated with advanced technologies such as AI and cloud computing, SCADA becomes a powerful tool for ensuring the reliability, efficiency and scalability of modern RES.

Key functions are as follows:

– real-time data collection and visualization;

– remote control of assets;

– alarm and fault management.

Sr. No.	SCADA feature/tool	Application/system considered	Reference
1.	Real-time data acquisition	Wind farm monitoring	Bentarzi et al. 2018
2.	HMI dashboards	Operator control in hybrid systems	Eltamaly et al. 2021
3.	Alarm management systems	Grid fault detection and alerting	Bentarzi et al. 2018
4.	PLC-SCADA interface	Substation automation	Deitsch et al. 2019
5.	Cloud-SCADA integration	Web-based RES asset monitoring	Bentarzi et al. 2018

Table 7.9. *SCADA and remote monitoring systems*

7.3.7. *Cloud and edge computing: distributed intelligence for energy applications*

Cloud and edge computing play a critical role in enhancing the digital infrastructure of RES. Cloud computing enables centralized data storage, scalable computing resources and advanced analytics by aggregating data from geographically distributed assets into central platforms. This allows for long-term performance tracking, predictive modeling and coordinated control strategies as in Table 7.10. In contrast, edge computing processes data locally at or near the point of generation such as within wind turbines, inverters or BMS. By executing key operations on-site, the edge computing significantly reduces data transmission latency and allows for real-time decision-making even in remote or bandwidth-constrained environments. The combination of cloud and edge computing ensures both the depth and speed of analysis by enabling continuous, resilient and intelligent energy system operations.

Key functions are as follows:

– scalability for data-heavy applications;

– fast, local decision-making;

– enhanced cybersecurity and data backup.

Sr. No.	Approach/platform	Application/system considered	Reference
1.	Edge analytics	Real-time decision-making at RES site	Liao et al. 2022
2.	Cloud-based data lakes	Centralized RES performance analytics	Liao et al. 2022
3.	Fog computing	Latency reduction in solar forecasting	Voyant et al. 2017
4.	AWS IoT/Azure IoT hub	Scalable data handling from sensor networks	Hameed et al. 2020
5.	Hybrid edge–cloud architecture	Efficient control in distributed energy systems	Liao et al. 2022

Table 7.10. *Cloud and edge computing*

7.4. Applications across renewable sources

7.4.1. *Application in solar energy*

7.4.1.1. *Solar forecasting*

Solar forecasting is crucial for predicting the future availability of solar irradiance and subsequent energy output from PV systems. Accurate forecasting enables better grid management, energy dispatch and battery scheduling as solar energy is inherently variable due to weather conditions. ML models such as long short-term memory (LSTM) neural networks have been increasingly employed to improve the precision of short-term and long-term solar radiation predictions as shown in Figure 7.7.

According to Voyant et al. (2017), ML methods significantly enhance the accuracy of solar radiation forecasting by providing a valuable tool for integrating solar energy into the grid more reliably and efficiently. A comparative summary of forecasting models for solar applications is shown in Table 7.11.

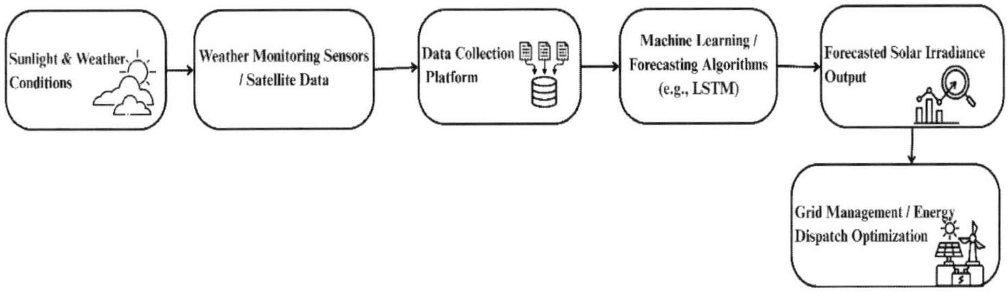

Figure 7.7. *Block diagram of solar forecasting using ML*

Sr. No.	Prediction time horizon	Application	Approach used	Reported accuracy	Reference
1.	Short-term 1–6 h	Real-time grid integration	LSTM neural network	RMSE:65–80 W/m²	Qing and Niu 2018
2.	Very short-term (15–60 min)	PV power smoothing and battery scheduling	CNN + LSTM hybrid	MAPE:3.5–6%	Zhu et al. 2021
3.	Day-ahead (24 h)	Load matching and energy market planning	Random forest + ARIMA hybrid	RMSE:120–135 W/m²	Zhu et al. 2021
4.	Short-term 1–12 h	Smart microgrid control	XGBoost regression	R^2:0.92, RMSE:75W/m²	Banik and Biswas 2024
5.	Intra-hour (5–15 min)	PV output ramp detection	ConvLSTM (spatiotemporal model)	MAE:28W/m²	Zang et al. 2024

Table 7.11. *Comparative summary of forecasting models for solar applications*

7.4.1.2. *Maximum power point tracking (MPPT) optimization*

MPPT optimization is a fundamental technique that ensures solar PV systems extract the maximum possible power under varying environmental conditions. MPPT algorithms dynamically adjust the operating voltage and current to find the optimal point at which the system produces maximum power. While traditional MPPT methods such as perturb and observe (P&O) and incremental conductance (INC) are widely used, modern AI-based approaches such as fuzzy logic controllers (FLC) and artificial neural networks (ANN) offer faster convergence and better adaptability especially under partial shading conditions as shown in Figure 7.8. Esram and Chapman (2007) highlighted that incorporating intelligent control strategies greatly improves MPPT efficiency, leading to higher overall system performance.

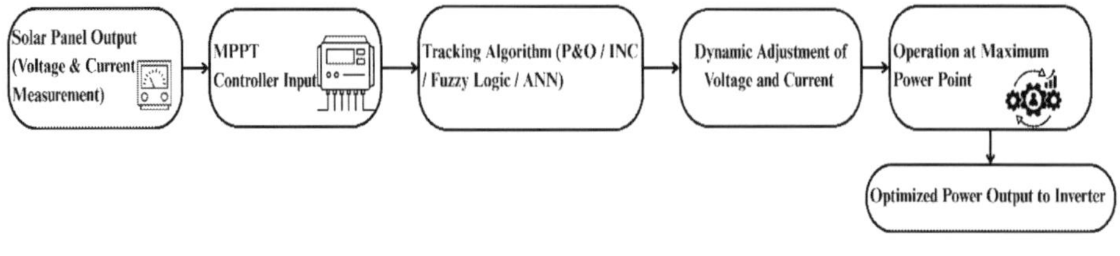

Figure 7.8. *Block diagram of MPPT optimization using AI*

7.4.1.3. *Inverter control*

Inverter control plays a crucial role in solar energy systems by converting DC generated by PV panels into AC suitable for residential, commercial and grid applications. Beyond basic conversion, modern smart inverters manage voltage and frequency stability, provide reactive power support and enhance grid resilience. Advanced control strategies including model predictive control (MPC) and AI-driven adaptive control allow inverters to respond effectively to dynamic grid conditions and support the integration of distributed energy resources, as shown in Figure 7.9. As described by Lopes et al. (2007), intelligent inverter control is a key enabler for the smooth integration of distributed generation into conventional power networks.

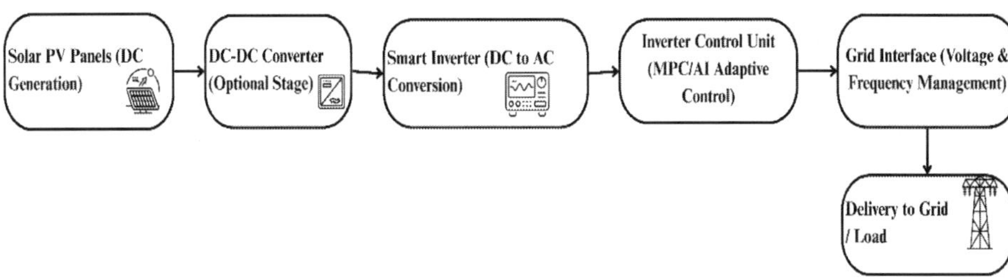

Figure 7.9. *Block diagram of inverter control using AI*

7.4.1.4. *AI for soiling, shading and fault detection*

AI has become a powerful tool for diagnosing and managing soiling, shading and faults in solar PV systems. Soiling from dust, partial shading from trees or buildings and internal module faults such as microcracks or hot spots can significantly degrade the performance of a PV module. AI techniques including deep learning models such as convolutional neural networks (CNNs) are employed to automatically detect and classify such issues based on images or sensor data as shown in Figure 7.10. Deitsch et al. (2019) demonstrated the effectiveness of AI models in automatically classifying defective PV cells using electroluminescence imaging, thus enabling proactive maintenance, minimizing downtime and optimizing energy yield.

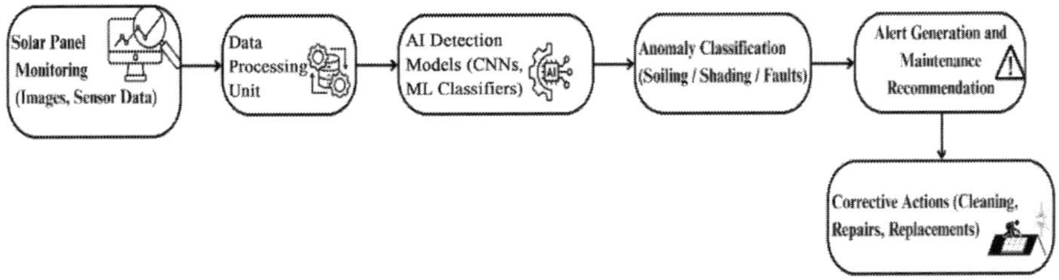

Figure 7.10. *Block diagram of using AI for soiling, shading and fault detection*

7.4.2. *Application in wind energy*

7.4.2.1. *AI for turbine health*

Wind turbines operate under highly variable environmental conditions, leading to wear and tear in the main components such as blades, gearboxes, generators and bearings. Traditional maintenance methods often rely on periodic inspections or reactive repairs after failure resulting in costly downtime. AI offers a smarter approach by enabling real-time turbine health monitoring through predictive analytics. ML models analyse large volumes of operational data such as vibration signals, temperature readings, rotational speed and power output to detect early signs of component degradation or faults. Techniques like support vector machines (SVM), random forests (RFs) and deep learning (e.g. CNN and LSTM networks) have been successfully applied to wind turbine fault diagnosis and health prediction as shown in Figure 7.11. According to Kusiak and Verma (2011), data-driven AI models significantly enhance the ability to predict status patterns and detect faults in wind turbines allowing early intervention and optimized maintenance strategies. Thus, AI-based health monitoring minimizes downtime, reduces operational costs and improves overall system efficiency.

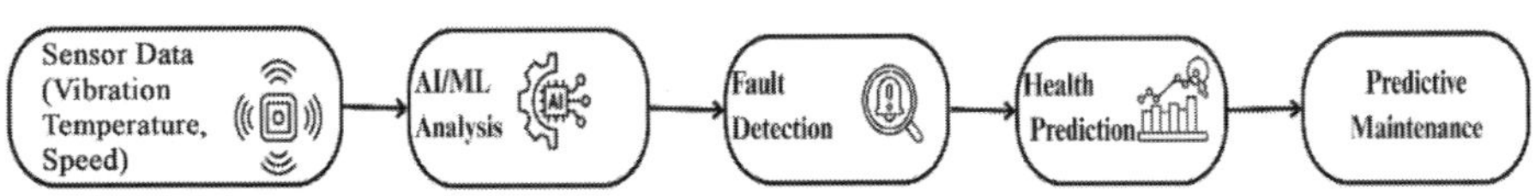

Figure 7.11. *Block diagram of turbine health using AI*

7.4.2.2. *AI for wind speed forecasting*

Wind speed is highly dynamic and difficult to predict accurately, yet precise forecasting is crucial for optimizing wind energy generation and maintaining grid stability. AI and ML models provide robust techniques to improve short-term and long-term wind speed forecasting by learning complex, nonlinear relationships from historical and real-time data. Methods such as ANNs, SVMs and RF are commonly used to forecast wind speeds based on variables like historical wind data, temperature, pressure and humidity as shown in Figure 7.12. Foley et al. (2012) highlight that AI-based forecasting models outperform traditional statistical methods, offering more reliable wind predictions essential for turbine control optimization, grid load management and energy trading. With better wind speed predictions, energy companies can improve operational efficiency, enhance financial planning and reduce the risks associated with wind variability.

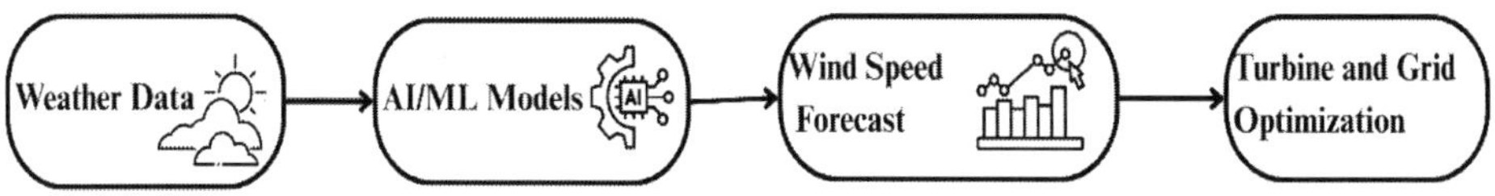

Figure 7.12. *Block diagram of wind speed forecasting using AI*

7.4.2.3. *Digital twins for predictive maintenance*

A digital twin is a virtual replica of a physical wind turbine that continuously updates in real-time using live sensor data. It simulates the behavior, operation and condition of the turbine under various operating scenarios, environmental conditions and load profiles. By comparing the real-world performance of the turbine with the simulated digital model, operators can identify deviations that may indicate developing faults, aging components or operational inefficiencies as shown in Figure 7.13. According to Rasheed et al. (2020), digital twins enable advanced predictive maintenance strategies by forecasting when and where failures are likely to occur based on data-driven models allowing repairs to be scheduled proactively rather than reactively. This approach reduces unplanned downtime, optimizes maintenance costs and extends the operational life of wind turbines ultimately leading to higher system reliability and energy production efficiency.

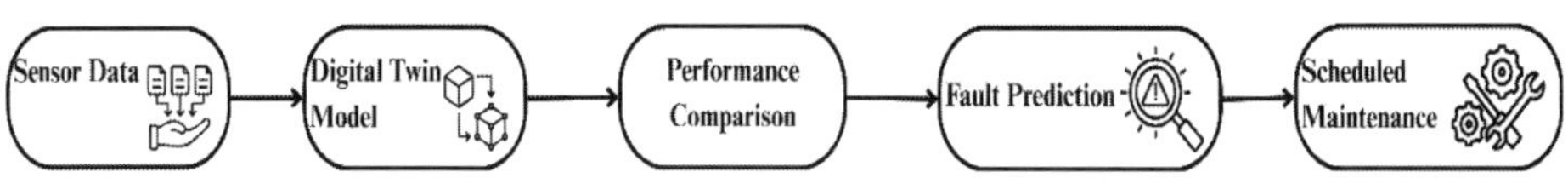

Figure 7.13. *Block diagram of digital twins for predictive maintenance*

7.4.3. *Application in hydro-electric energy*

7.4.3.1. *Flow prediction and dam safety automation*

Flow prediction and dam safety automation are two of the most crucial applications of digital technology in hydro energy systems. Accurate flow prediction ensures optimal reservoir management, efficient power generation scheduling and effective flood risk mitigation. Modern approaches leverage ML algorithms, such as ANNs, SVMs and RF models, and trained with historical weather and hydrological data to predict river inflows accurately, as shown in Figure 7.14. Additionally, remote sensing technology combined with GIS allows continuous monitoring of watershed and snowpack conditions, which significantly improve prediction models. Real-time data acquisition through IoT-enabled sensor networks provides critical inputs like rainfall intensity, soil moisture levels and river stage heights, forming the backbone of intelligent forecasting systems.

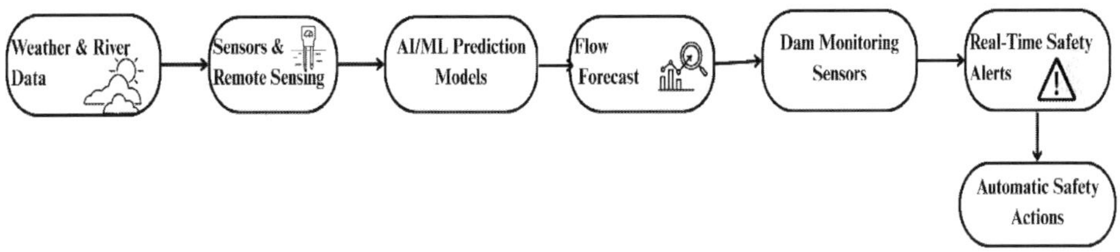

Figure 7.14. *Block diagram of flow prediction and dam safety automation using AI*

Dam safety automation relies heavily on real-time monitoring and predictive analytics to maintain structural integrity and operational security. Advanced structural health monitoring (SHM) systems employ sensor networks to measure parameters such as stress, seepage and vibration, ensuring early detection of potential risks. Early warning systems (EWS) powered by AI analyze this sensor data issuing alerts when anomalies are detected. The concept of digital twins' virtual replicas of dam infrastructure updated in real time enables predictive simulations and scenario planning, ensuring that dam operators can proactively manage risks and adhere to regulatory standards. Through these digital advancements, dam safety has moved toward a more proactive and preventive framework.

7.4.3.2. *SCADA in hydro plant operation*

SCADA systems play an integral role in the efficient operation of hydroelectric plants by enabling centralized supervision, monitoring and control. These systems are designed to collect, process and present operational data from across the plant, supporting real-time decision-making and enhancing overall plant performance. The typical SCADA architecture includes remote terminal units (RTUs) that gather field data from turbines, gates and spillways the programmable logic controllers (PLCs) that execute automated control logic and human–machine interfaces (HMIs) that display real-time system status to operators. Communication between these components is facilitated by robust infrastructure such as fiber optics, radio or satellite links, which ensure seamless and reliable data flow as shown in Figure 7.15.

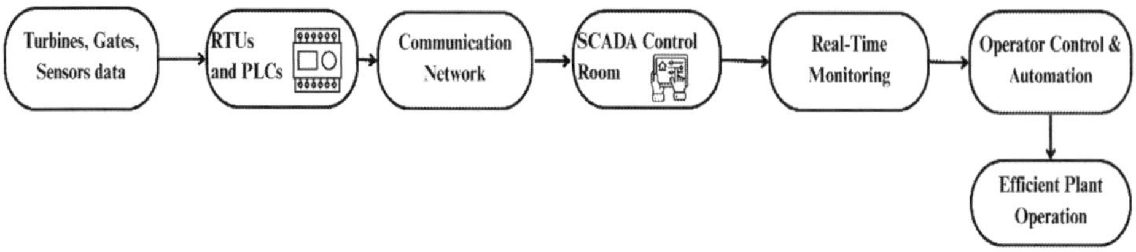

Figure 7.15. *Block diagram of SCADA in hydro plant operation*

Key functionalities enabled by SCADA systems include real-time monitoring of critical parameters like water levels and turbine performance, automated control of mechanical components based on operational conditions and predictive maintenance by identifying equipment degradation trends. SCADA also supports data analytics and historical data management, empowering operators to optimize energy production and predict future performance trends. Alarm management systems integrated into SCADA promptly alert operators to critical events, which minimize response time during abnormal situations. The integration of SCADA into hydro plant operations offers significant advantages such as enhanced operational efficiency, reduced human error, improved cybersecurity and better-informed decision-making processes, which contribute to the resilience and sustainability of hydro energy infrastructure.

7.4.4. *Application in biomass/biogas*

7.4.4.1. *AI for feedstock prediction and combustion control*

AI is increasingly used in biomass and biogas plants to enhance feedstock management and combustion efficiency. For feedstock prediction, AI models analyze multiple variables such as weather data, crop yields, harvesting schedules and historical supply patterns to forecast the availability and quality of biomass inputs. This predictive approach enables operators to plan better, optimize storage and maintain a steady and reliable feedstock supply. In combustion control, AI algorithms monitor and adjust operational parameters such as fuel feed rate, air-to-fuel ratio, temperature and pressure in real-time, as shown in Figure 7.16. By continuously learning from operational data, AI ensures that combustion occurs under optimal conditions, maximizing energy output while minimizing emissions such as carbon monoxide and unburned hydrocarbons. This not only boosts plant efficiency but also supports compliance with stringent environmental standards.

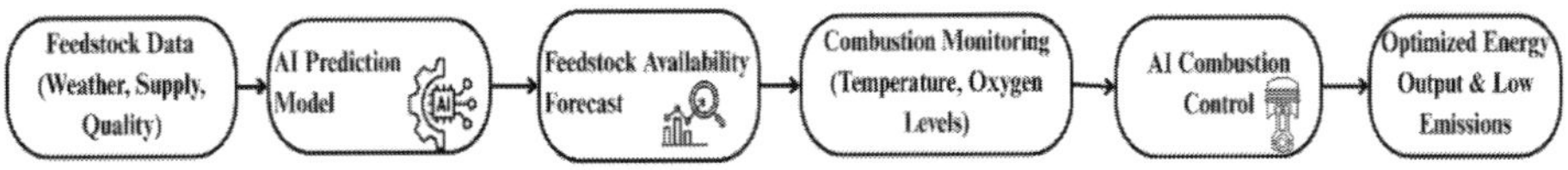

Figure 7.16. *Block diagram of AI for feedstock prediction and combustion control*

7.4.4.2. *Remote plant management and emissions monitoring*

Remote plant management in biomass and biogas facilities has been revolutionized by the integration of IoT sensors, cloud computing and advanced communication technologies. Operators can now oversee plant operations, track equipment performance and execute control commands from remote locations using centralized platforms. Real-time data visualization and alarm systems allow for quicker detection of faults, improving response times and reducing maintenance costs. Alongside operational monitoring, digital solutions have greatly enhanced emissions monitoring. Sensors installed throughout the plant continuously measure levels of greenhouse gases, particulate matter and other pollutants. The data are automatically logged, analyzed and compared against regulatory thresholds. Immediate alerts are triggered if emissions exceed permitted levels, ensuring swift corrective actions, as shown in Figure 7.17. Overall, remote management combined with automated emissions monitoring leads to improved operational efficiency, environmental compliance and sustainable plant performance.

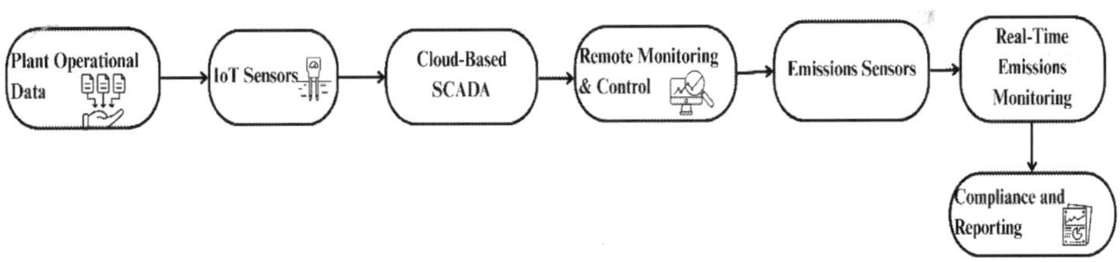

Figure 7.17. *Block diagram of remote plant management and emissions monitoring using IOT*

7.4.5. *Application in hybrid and microgrid systems*

7.4.5.1. *Energy management systems*

In hybrid and microgrid systems, digital technologies have revolutionized EMS enabling real-time optimization, automation and predictive control. Digital EMS platforms use sensors, IoT devices, advanced controllers and cloud computing to monitor the entire energy ecosystem including solar panels, wind turbines, batteries, diesel generators and loads. AI and ML algorithms are integrated into EMS to forecast energy generation from renewables, predict load demand and suggest optimal dispatch strategies, as shown in Figure 7.18. These digital EMS platforms continuously adjust operations by controlling energy flows, minimizing operational costs, maximizing the use of renewable energy and improving battery lifecycle management. Cloud-based EMS solutions further allow remote monitoring, remote control, data analytics and real-time decision-making by enabling microgrids to operate more autonomously and resiliently, as presented in Table 7.12. Cybersecurity tools are also embedded to protect EMS from cyberattacks, ensuring safe and uninterrupted operation.

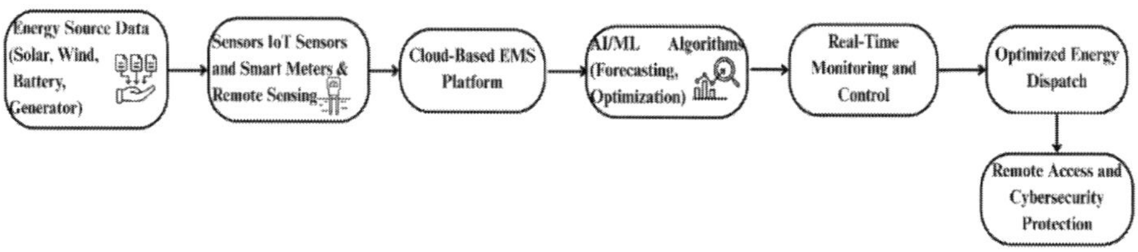

Figure 7.18. *Block diagram of energy management systems (EMSs) using IOT and AI*

Sr. No.	EMS approach/component	Application/feature	Reported benefit/metric	Reference
1.	AI-based load and generation forecasting	Optimized energy dispatch in solar–diesel–battery hybrid	15 – 20 % reduction in fuel use	Banik and Biswas 2024
2.	IoT-integrated EMS	Real-time monitoring & energy flow control	Latency < 5 sec for data sync	Hameed et al. 2020
3.	Cloud-based EMS platform	Remote EMS control and diagnostics	24/7 accessibility, data logs	IRENA (2022)
4.	Adaptive EMS with reinforcement learning	Optimal source switching in microgrids	Cost reduction: 18 %, RES use: +22 %	Lee and Choi 2019
5.	Cybersecurity-integrated EMS	Secure data flow and control	Zero downtime from attacks (case)	Zhang et al. (2024)

Table 7.12. *Digital EMS approaches in hybrid systems*

7.4.5.2. *Load prediction and automated switching*

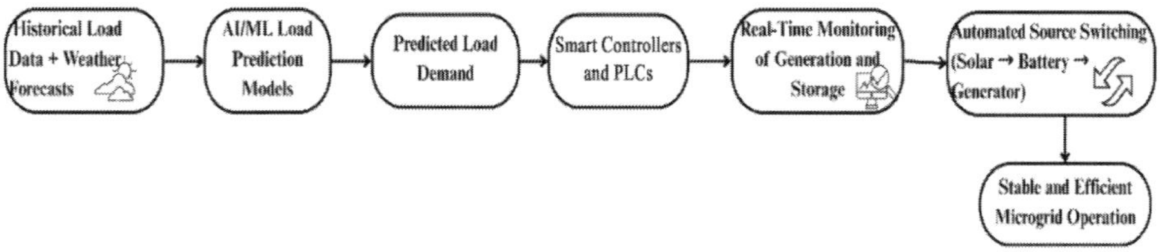

Figure 7.19. *Block diagram of load prediction and automated switching using AI*

Sr. No.	Approach/technology	Application in hybrid system	Accuracy/outcome	Reference
1.	LSTM neural network	Short-term load prediction	Mean absolute percentage error (MAPE): 4.8%	Zhu et al. 2021
2.	XGBoost + feature selection	Hourly demand forecasting for smart EMS	R^2: 0.94	Banik and Biswas 2024
3.	Smart inverter + PLC controller	Automated source switching (solar/battery/gen)	Switch delay < 2 s	Eltamaly et al. 2021
4.	Event-based forecasting (AI)	Dynamic load behavior modeling (special events)	Mean absolute percentage (MAP): 3.2% during peak shifts	Banik and Biswas 2024
5.	Edge controller + remote cloud	Autonomous load balancing	99.5% uptime in islanded mode	Liao et al. 2022

Table 7.13. *Comparative summary: load forecasting and switching approaches*

Digital applications have significantly improved load prediction and automated switching functionalities in hybrid and microgrid systems. Advanced load prediction now uses ML models trained on historical consumption patterns, weather forecasts, time-of-day behaviors and event-based factors to accurately predict upcoming electricity demand. These AI-driven models continuously learn and improve by adapting to changing user behaviors and environmental conditions, as shown in Figure 7.19. For automated switching, digital controllers and intelligent relay systems enable seamless switching between different energy sources (e.g. from solar to battery to generator) without manual intervention. Smart inverters by PLCs and real-time control software execute switching operations based on predefined algorithms and real-time conditions. Automation ensures that the most efficient and sustainable energy source is prioritized while maintaining system stability, as presented in Table 7.13. Cloud-based digital platforms allow remote supervision of switching activities and quick adaptation to unexpected load changes or generation variability. Combined, digital load prediction and automated switching increase the reliability, efficiency and sustainability of hybrid and microgrid operations.

7.5. Predictive maintenance and asset management

Predictive maintenance and asset management are critical for enhancing the efficiency, reliability and lifespan of RES such as solar farms and wind farms. Unlike traditional maintenance methods, predictive maintenance uses real-time sensor data and advanced analytics powered by AI and ML to detect early signs of equipment degradation or anomalies. By forecasting potential failures before they occur, the operators can schedule timely interventions that reduce unplanned downtime, optimize maintenance costs and extend component life. Asset management complements this approach by systematically tracking the performance, condition and financial metrics of each asset throughout its lifecycle by ensuring maximum value extraction. Together, predictive maintenance and asset management not only lower operational risks and costs but also improve energy output, support grid stability and contribute to the long-term sustainability of renewable energy infrastructures.

7.5.1. *Role of AI and ML in anomaly detection*

AI and ML have transformed predictive maintenance in RES by enabling advanced anomaly detection. In solar and wind farms, vast amounts of operational data are continuously generated, including temperature, vibration, voltage, current, rotational speed and power output measurements. AI and ML algorithms such as SVM, RF, neural networks and autoencoders (AE) are trained on historical and real-time data to learn the normal behavior of assets. These models can identify subtle deviations from normal patterns, detecting early signs of faults, component degradation or inefficiencies that would be difficult for human operators to recognize. Through anomaly detection, the predictive maintenance becomes proactive rather than reactive, allowing early interventions that prevent minor issues from escalating into major failures. This not only reduces downtime and repair costs but also improves the overall reliability and lifespan of renewable energy assets.

7.5.1.1. *AI-Powered predictive maintenance in a 75 MW solar farm (Arizona, USA)*

A comprehensive case study (Sarkar 2024) details the implementation of an AI-based predictive maintenance system at Solar Tech Solutions' 75 MW solar installation in Arizona. Over an 18-month period, the integration of 12,000 IoT sensors with advanced AI/ML infrastructure led to significant operational improvements:

– 94.3% accuracy in anomaly detection;

– 98.2% precision in fault localization;

– 47% reduction in unplanned downtime, equating to annual savings of $425,000;

– 64% increase in mean time between failures (MTBF);

– reduction in maintenance response time from 72 to 4 h;

– 3.2% improvement in panel efficiency;

– annual reduction of 1,960 metric tons of CO_2 emissions;

– conservation of 1.2 million gallons of water per year.

Sr. No.	Company/platform	Application area	Key features/outcomes	Reference
1.	Siemens Energy – *Simcenter*	Wind farm performance optimization	Real-time monitoring, predictive failure detection and control tuning for turbine blades and gearboxes	Potter 2025
2.	GE Renewable Energy – *Digital Wind Farm*	Wind energy forecasting and asset management	Increased annual energy production (AEP) by up to 20% via predictive analytics	GE Renewable Energy (2023)
3.	DNV – *Twinify*	Offshore wind farms	Lifetime structural modeling, fatigue tracking and risk analytics	Ørsted 2021
4.	ABB Ability Energy Manager	Solar + Battery Microgrids	Cloud-based twin for real-time EMS optimization and fault diagnostics	ABB (2023)
5.	Ansys Twin Builder	Solar inverter digital modeling	Simulation of inverter control under different irradiance and temperature conditions	Ansys (2023)
6.	Microsoft Azure + Schneider Electric – *EcoStruxure*	Microgrids (Campus + Industrial)	Predictive control and self-healing of grid-connected systems using digital twin models and cloud edge AI	Schneider Electric United States, 2024

Table 7.14. *Digital twin platforms in renewable energy*

This case study exemplifies the transformative impact of AI-driven maintenance solutions in utility-scale solar operations, highlighting both economic and environmental benefits. Digital twin platforms in renewable energy are shown in Table 7.14.

7.5.2. *Preventing failures and extending component life*

Predictive maintenance driven by digital technologies focuses on preventing failures and extending the operational life of critical components in solar panels, wind turbines, inverters, gearboxes and electrical systems. By continuously monitoring the condition of equipment and predicting potential failures, the maintenance actions can be scheduled precisely, which is required to avoid unnecessary inspections and minimizing disruptions. For instance, in wind turbines, the predictive models monitor gearboxes for vibration anomalies that signal early-stage wear; in solar farms, the thermal imaging and electrical data help to predict hotspot formation and panel degradation. Targeted maintenance interventions address specific issues before they cause breakdowns, thus preserving asset health. Additionally, data-driven insights from AI can recommend operational adjustments (e.g. load balancing and reduced operating stress) to extend the service life of components. This approach leads to reduced lifecycle costs, optimized performance, increased energy yield and a more sustainable energy infrastructure.

7.5.3. *Case studies across solar and wind farms*

Several case studies demonstrate the practical success of predictive maintenance and asset management in solar and wind energy sectors. In one notable example, a wind farm implemented AI-based vibration monitoring and anomaly detection, which identified subtle gearbox issues months before a major failure was likely to occur, allowing preventive repairs and saving substantial costs. Similarly, in a large solar farm, thermal imaging combined with ML models helps detect early-stage module degradation and inverter faults, leading to timely maintenance actions and significant improvements in system efficiency. Another study highlighted how a combination of digital twins and predictive analytics in offshore wind farms led to a 20% reduction in maintenance costs and a measurable increase in turbine uptime. These real-world cases show that digital predictive maintenance not only enhances the reliability of renewable energy assets but also improves financial performance, energy output and operational sustainability.

7.5.3.1. *Predictive maintenance for offshore wind turbines using deep learning*

Predictive maintenance using DL techniques has revolutionized the operation and management of offshore wind turbines. These turbines are deployed in challenging marine environments characterized by high humidity, saltwater exposure, strong winds and variable mechanical loads. Such extreme conditions accelerate the degradation of critical components, including blades, gearboxes, generators and tower structures. Traditionally, offshore wind turbine maintenance has been based on periodic inspections or reactive repairs. However, these conventional approaches are often costly and ineffective in preventing unexpected failures, leading to increased downtime and operational risk. To address these limitations, a proactive predictive maintenance strategy using DL has gained prominence. By analysing real-time sensor data such as vibration signals, temperature measurements and rotational speed patterns, the DL models can detect early signs of faults and predict failures before they occur. This data-driven approach facilitates early interventions, thereby reducing unplanned outages, minimizing maintenance costs and extending the lifespan of turbine components.

Two widely adopted DL models in this context are:

1) CNNs: which are ideal for extracting spatial features and patterns from complex sensor data.

2) Recurrent neural networks (RNNs): which are capable of modeling time-dependent behaviors and identifying trends in sequential data.

These models automatically learn nonlinear relationships and subtle patterns associated with system degradation patterns that traditional techniques often fail to capture. As a result, they enable enhanced:

– early fault detection;

– anomaly diagnosis;

– failure forecasting.

While CNNs and RNNs are frequently used due to their strengths in spatial and temporal pattern recognition, respectively, several other DL models have shown competitive or superior performance in recent studies.

Other reported approaches

– *AE*: are used for anomaly detection by reconstructing baseline operational behavior and flagging deviations.

– *LSTM*: is a variant of RNNs that handles long-term dependencies, effective for sequential turbine sensor data.

– *Transformer-based models*: recent advances show that transformer architectures outperform LSTM in multi-sensor time-series analysis for turbine fault prediction.

– *Deep belief networks (DBN)*: are applied for unsupervised pretraining and classification of turbine failure modes.

– *Hybrid models*: these are used for combining ML models (e.g. RFs or SVMs) with deep feature extraction layers from CNNs or AEs for improved robustness.

Downtime reduction through digital predictive maintenance

Digital predictive maintenance systems significantly reduce mean time to repair (MTTR) and unscheduled downtime by:

– detecting faults weeks before failure occurs (Udo et al. 2023);

– increasing availability by 5–15% (Al-Shetwi et al. 2025);

– reducing O&M costs by up to 25% by enabling condition-based interventions (Rasheed et al. 2019).

These improvements are particularly critical for offshore wind, where logistical and weather constraints make maintenance access challenging and expensive.

Recent studies have further demonstrated the impact of integrating deep learning with online clustering techniques to improve fault prediction.

For example, Pineda et al. (2024) presented a real-world case study where DL combined with online clustering successfully predicted failures in unsupervised subsystems of offshore wind turbines. This hybrid approach was especially effective in environments where labelled data for faults was limited, yet operational data were abundant. The application of such methods leads to numerous operational advantages, including improved maintenance planning, reduced downtime, increased reliability, cost savings and longer equipment lifespan.

In summary, DL–based predictive maintenance enables a shift from reactive to intelligent data-driven strategies in offshore wind operations. By enhancing system awareness and enabling timely fault intervention, it ensures more resilient, efficient and economically viable management of offshore wind energy assets.

7.6. Data-driven decision-making

Data-driven decision-making is transforming RES by using real-time and historical data to optimize operations. By integrating weather forecasts, market pricing and grid status, operators can predict energy availability, respond to market opportunities and dispatch power more effectively. Load forecasting powered by AI and ML helps anticipate electricity demand with high accuracy by enabling better planning and resource allocation. Meanwhile, smart meters play a key role in DSM by providing detailed consumption data, allowing utilities to adjust pricing, reduce peak load and enhance grid stability. Together, these digital tools support smarter, more responsive and efficient energy systems. Differences in the data-driven approaches discussed are presented in Table 7.15.

Sr. No.	Approach	Purpose	Input data	Role in RES
1.	AI/ML-based load forecasting	Predict electricity demand accurately	Historical consumption, weather, time-of-day data	Enables better scheduling and energy resource allocation
2.	Weather forecast integration	Anticipate renewable energy availability (solar/wind)	Real-time and forecasted meteorological data	Supports generation planning and dispatch decisions
3.	Market price data analysis	Optimize dispatch based on market conditions	Electricity pricing, tariffs, market signals	Maximizes economic return, reduces cost of energy use
4.	Smart meter analytics	Improve demand-side management and consumer profiling	High-resolution electricity consumption data	Enables dynamic pricing, peak load reduction
5.	Grid status feedback monitoring	Maintain grid balance and detect operational issues	Voltage, frequency, load flow, fault data	Enhances grid stability and operational responsiveness

Table 7.15. *Data-driven approaches*

7.7. Case studies

7.7.1. *Enel's digital wind farms*

Enel Group (2020), a global energy company, has implemented advanced digital technologies in its wind farms, using sensors, IoT and AI-based analytics to monitor turbine performance in real-time. This has enabled predictive maintenance, reduced downtime and improved energy output, setting a strong example of how digital transformation enhances operational efficiency in wind energy.

7.7.2. *Google DeepMind in wind optimization*

Google DeepMind (2019) used ML to optimize energy production from wind farms by predicting wind output 36 h in advance. By feeding these forecasts into scheduling systems, they increased the value of wind energy by about 20%, showcasing how AI can improve forecast accuracy and grid integration.

7.7.3. *India's national solar mission SCADA integration*

MNRE (2021), Under India's National Solar Mission, SCADA systems were deployed across large-scale solar parks to enable centralized monitoring and control. This integration allows for real-time data collection on generation, weather and equipment status, leading to better asset management and grid stability in one of the world's fastest-growing solar markets.

7.7.4. *Blockchain pilots in Germany and Australia for peer energy trading*

Lu et al. 2019, pilot projects in Germany and Australia have tested blockchain platforms that allow consumers to trade excess solar energy directly with neighbors. These decentralized systems use smart contracts and transparent ledgers to enable secure, real-time P2P energy trading, promoting local energy markets and grid resilience.

7.8. Challenges and considerations

While digitalization offers significant advantages for RES, it also introduces several challenges that must be carefully addressed. One major concern is data privacy and cybersecurity as increased connectivity and real-time data transmission expose energy systems to potential cyber threats and data breaches. Protecting sensitive operational and consumer information is critical to maintaining trust and system integrity. Interoperability and standardization present another hurdle as diverse digital devices, platforms and communication protocols must seamlessly work together without common standards, and system integration becomes complex and inefficient. Additionally, the high initial investment required for deploying digital infrastructure, including sensors, IoT devices, AI platforms and cloud services, can be a significant barrier, especially for small- and medium-sized energy providers. Finally, capacity building and skill development are essential as operating and maintaining advanced digital systems requires a workforce skilled in data analytics, cybersecurity, AI technologies and energy systems engineering. Addressing these challenges is vital to fully realizing the benefits of digital transformation in the renewable energy sector.

7.8.1. *Cost of digitalization*

One of the most quantifiable challenges is the initial capital cost associated with digital infrastructure. For instance, deploying EMS with IoT-based metering, edge controllers, AI-driven forecasting modules and cloud-based dashboards for a small-scale 100 kW hybrid microgrid system can incur costs in the range of ₹15–20 lakhs ($18,000–$24,000). This includes:

– IoT sensors and gateways: ₹2–3 lakhs;

– edge computing units and PLCs: ₹3–4 lakhs;

– AI/ML software licenses or development: ₹ 3–5 lakhs;

– cloud integration and security: ₹ 2–3 lakhs;

– skilled manpower and training: ₹2–5 lakhs.

Such costs are often prohibitive for small and medium RES operators, especially in rural or islanded systems, making scalability and affordability key concerns.

7.8.2. *Model development and deployment challenges*

Developing robust AI/ML models for applications like load forecasting, fault detection or predictive maintenance requires:

– large volumes of high-quality, labeled data, which are often unavailable or inconsistent in RES setups;

– domain-specific feature engineering, as raw data from solar PV, wind and storage systems differ in behavior and resolution;

– computational complexity, where deep learning models (e.g. LSTM and Transformer) need powerful GPUs and extended training cycles;

– deployment bottlenecks, including the integration of trained models into live EMS platforms, real-time decision execution and compatibility with legacy systems;

– model drift and retraining needs, as environmental and load conditions evolve over time.

7.8.3. *Cybersecurity and data privacy*

Increased connectivity through IoT and cloud systems creates new cyberattack surfaces. Threats such as data spoofing, denial of service (DoS), ransomware and unauthorized access can compromise system operation or steal sensitive data. Despite availability of solutions like blockchain and edge security, cybersecurity remains underfunded in small-scale deployments. A breach in control systems can lead to:

– unsafe grid operations;

– cascading failures in hybrid setups;

– violation of consumer privacy regulations (e.g. GDPR).

7.8.4. *Interoperability and standardization*

Digital RES systems often incorporate multi-vendor equipment using different communication protocols (e.g. Modbus, DNP3 and MQTT). The lack of universal standards hinders seamless data exchange and automation. Without interoperability frameworks like IEC 61850 or IEEE 2030.5, integrating digital twins, SCADA, AI modules and EMS into a unified platform becomes complex and time consuming.

7.8.5. *Human capital and skill gaps*

Digital RES systems require operators skilled in:

– energy systems modeling;

– data science and AI;

– cybersecurity architecture;

– cloud and edge deployment.

There is currently a critical gap in workforce capabilities in these domains, especially in developing countries. Without structured training programs and upskilling, utilities may struggle to fully exploit the value of digitalization.

7.9. Future directions

The future of renewable energy is set to be increasingly autonomous and digitally empowered. Autonomous energy systems, driven by AI and advanced control algorithms, will allow for real-time self-optimization of generation, storage and consumption without human intervention. In parallel, the rise of digital prosumers (consumers who also produce and manage their own energy) will create more decentralized and interactive energy markets. Federated learning is a decentralized ML approach that allows AI models to be trained across multiple edge devices or servers holding local data sample without exchanging the actual data, and decentralized AI models are emerging to address data privacy concerns by enabling AI training across distributed energy resources without centralizing sensitive data. This approach will enhance collaborative intelligence while preserving security. Furthermore, AI applications are expanding into circular energy systems and waste-to-energy technologies, optimizing resource recovery, enhancing sustainability and reducing environmental impact. Lastly, the rapid pace of digitalization demands supportive policy frameworks that encourage innovation while ensuring cybersecurity, interoperability and equitable access. Well-structured regulations will be essential for guiding responsible growth and enabling a resilient, digitally driven energy future.

7.9.1. *Emerging digital trends*

– *Autonomous energy systems*: advanced AI-based controllers and multi-agent systems will enable fully autonomous operations optimizing generation, storage and dispatch in real-time without human intervention.

– *Digital prosumers and energy communities*: the emergence of prosumers and P2P trading requires secure digital platforms and decentralized market algorithms to ensure fairness, efficiency and privacy.

– *Federated learning and edge AI*: to address data privacy and latency challenges, federated learning enables AI models to be trained across distributed devices without moving data to a central server. This is vital for decentralized microgrids and rural energy systems.

– *Circular and waste-to-energy systems*: AI and IoT tools are being adapted to optimize resource utilization in integrated waste-to-energy plants, supporting circular economy goals.

– *Policy and cyber governance*: the growth of digital RES demands new policy frameworks that prioritize interoperability, cybersecurity and equitable access to digital energy services.

7.9.2. *Opportunities for research*

Table 7.16 highlights emerging research opportunities in digital RESs, spanning forecasting, energy management, digital twins and smart grid analytics. By leveraging AI, blockchain and interoperability frameworks, these areas promise scalable, secure and sustainable solutions for the future energy landscape.

Application area	Research opportunity
Solar forecasting	Developing hybrid models (e.g. Transformer + LSTM) using multi-modal data (images, time-series and weather maps)
Energy management systems (EMSs)	Real-time EMS algorithms using reinforcement learning and federated learning for decentralized control
Load prediction and switching	Adaptive ML models that respond to dynamic sociobehavioral patterns and seasonal variability
Digital twins	Creating scalable, real-time twins for multi-source hybrid systems (solar-wind-storage-load), integrated with live SCADA data
Predictive maintenance	Self-learning anomaly detection models trained on limited or noisy data using few-shot or semi-supervised learning
Smart meter analytics	Privacy-preserving analytics using differential privacy and secure multi-party computation for demand profiling
Blockchain and P2P trading	Lightweight blockchain frameworks suited for energy-constrained edge devices in microgrids
Interoperability frameworks	Developing open-source middleware for IEC 61850/MQTT/IEEE 2030.5 protocol harmonization
Circular economy applications	AI models for optimizing thermal and biological waste-to-energy conversion pathways

Table 7.16. *Opportunities for research*

7.10. Conclusions

Digitalization is a powerful catalyst driving the transformation of RES toward greater efficiency, reliability and sustainability. By leveraging AI, IoT and data analytics, energy systems are becoming smarter and more adaptive. To fully realize this potential, strategic efforts are needed for policymakers to create supportive regulatory frameworks, industry leaders should invest in scalable digital infrastructure and researchers must continue advancing innovative, secure and interoperable technologies. Together, these actions will pave the way for a resilient and sustainable digital energy future.

7.11. References

ABB (2023). ABB Ability™ energy manager: Smart Power Digital Solutions [Online]. Available at: https://new.abb.com/low-voltage/products/smart-power-digital-solutions/abb-ability-energy-manager [Accessed 9 June 2025].

Aghenta, L.O. (2020). Opensource SCADA systems for small renewable power generation. PhD Thesis, Memorial University of Newfoundland, St. John's.

Al-Shetwi, A.Q., Atawi, I.E., El-Hameed, M.A., Abuelrub, A. (2025). Digital twin technology for renewable energy, smart grids, energy storage and vehicle-to-grid integration: Advancements, applications, key players, challenges and future perspectives in modernising sustainable grids. *IET Smart Grid*, 8(1), e70026. doi: 10.1049/STG2.70026.

Alarcon, A. (2019). The digital revolution of hydropower in Latin American countries. Inter-American Development Bank [Online]. Available at: https://publications.iadb.org/publications/english/document/The_Digital_Revolution_of_Hydropower_in_Latin_American_Countries_en.pdf [Accessed June 2025].

Ansys (2023). Ansys twin builder: Create and deploy digital twin models [Online]. Available at: https://www.ansys.com/products/digital-twin/ansys-twin-builder [Accessed 9 January 2026].

Arch, A., Cortijo, R., Romero, E., Canga, E., Furrer, P., Woodhouse, S., Dulle, H., Koller, T. (2019). *The Digital Revolution of Hydropower in Latin American Countries*. Inter-American Development Bank, Washington, D.C.

Banik, R. and Biswas, A. (2024). Enhanced renewable power and load forecasting using RF-XGBoost stacked ensemble. *Electrical Engineering*, 106(4), 4947-04967. doi: 10.1007/S00202-024-02273-3.

Bentarzi, H., Tsebia, M. and Abdelmoumene, A. (2018). PMU based SCADA enhancement in smart power grid. In *Proceedings – 2018 IEEE 12th International Conference on Compatibility, Power Electronics and Power Engineering (CPE-POWERENG 2018)*, 1–6. doi: 10.1109/CPE.2018.8372580.

Deitsch, S., Christlein, V., Berger, S., Buerhop-Lutz, C., Maier, A., Gallwitz, F., Riess, C. (2019). Automatic classification of defective photovoltaic module cells in electroluminescence images. *Solar Energy*, 185, 455–468. doi: 10.1016/J.SOLENER.2019.02.067.

Eltamaly, A.M., Alotaibi, M.A., Alolah, A.I., Ahmed, M.A. (2021). IoT-based hybrid renewable energy system for smart campus. *Sustainability*, 13(15), 8555. doi: 10.3390/SU13158555.

Enel Group (2020). Enel Green Power brings online two new U.S. wind farms, including its largest renewable plant in operation worldwide [Online]. Available at: https://www.enel.com/media/explore/search-press-releases/press/2020/12/enel-green-power-brings-online-two-new-us-wind-farms-including-its-largest-renewable-plant-in-operation-worldwide- [Accessed June 12, 2025].

Esram, T. and Chapman, P.L. (2007). Comparison of photovoltaic array maximum power point tracking techniques. *IEEE Transactions on Energy Conversion*, 22, 439–449. doi: 10.1109/TEC.2006.874230.

Foley, A.M., Leahy, P.G., Marvuglia, A., McKeogh, E.J. (2012). Current methods and advances in forecasting of wind power generation. *Renewable Energy*, 37(1), 1–8. doi: 10.1016/J.RENENE.2011.05.033.

Google DeepMind (2019). Machine learning can boost the value of wind energy [Online]. Available at: https://deepmind.google/blog/machine-learning-can-boost-the-value-of-wind-energy/ [Accessed 12 June 2025].

Hameed, Z., Ahmad, F., ur Rehman, S., Ghafoor, Z. (2020). IoT based communication technologies to integrate and maximize the efficiency of renewable energy resources with smart grid. In *2020 International Conference on Computing and Information Technology (ICCIT-1441)*. IEEE. doi: 10.1109/ICCIT-144147971.2020.9213730.

IEA Bioenergy (2021). The first bioenergy village in Jühnde/Germany: Energy self sufficiency with biogas. [Online] Available at: https://www.ieabioenergy.com/wp-content/uploads/2018/01/biogas_village.pdf

International Energy Agency (2023). Renewables 2023: Analysis and Forecast to 2028. Executive summary. Press Release, IEA, Paris [Online]. Available at: https://www.iea.org/reports/renewables-2023/executive-summary [Accessed 9 June 2025].

International Renewable Energy Agency (IRENA) (2022). *World Energy Transitions Outlook 2022: 1.5°C Pathway*. IRENA, Abu Dhabi [Online]. Available at: https://gws-os.com/en/publications/all-publications/detail/world-energy-transitions-outlook-2022 [Accessed 12 June 2025].

Jahangir, H., Tayarani, H., Gougheri, S.S., Golkar, M.A., Ahmadian, A., Elkamel, A. (2020). Deep learning-based forecasting approach in smart grids with microclustering and bidirectional LSTM network. *IEEE Transactions on Industrial Electronics*, 68(9), 8298–8309. doi: 10.1109/TIE.2020.3009604.

Kusiak, A. and Verma, A. (2011). Prediction of status patterns of wind turbines: A data-mining approach. *Journal of Solar Energy Engineering*, 133(1), 011008. doi: 10.1115/1.4003188/444215.

Lee, S. and Choi, D.H. (2019). Reinforcement learning-based energy management of smart home with rooftop solar photovoltaic system, energy storage system, and home appliances. *Sensors*, 19(18), 3937. doi: 10.3390/S19183937.

Liao, H., Zhou, Z., Liu, N., Zhang, Y., Xu, G., Wang, Z., Mumtaz, S. (2022). Cloud-edge-device collaborative reliable and communication-efficient digital twin for low-carbon electrical equipment management. *IEEE Transactions on Industrial Informatics*, 19(2), 1715–1724. doi: 10.1109/TII.2022.3194840.

Lopes, J.A.P., Hatziargyriou, N., Mutale, J., Djapic, P., Jenkins, N. (2007). Integrating distributed generation into electric power systems: A review of drivers, challenges and opportunities. *Electric Power Systems Research*, 77(9), 1189–1203. doi: 10.1016/J.EPSR.2006.08.016.

Lu, X., Guan, Z., Zhou, X., Du, X., Wu, L., Guizani, M. (2019). A secure and efficient renewable energy trading scheme based on blockchain. In *Proceedings of 21st IEEE International Conference on High Performance Computing and Communications, 17th IEEE International Conference on Smart City and 5th IEEE International Conference on Data Science and Systems*. HPCC/SmartCity/DSS, 1839–1844. doi: 10.1109/HPCC/SMARTCITY/DSS.2019.00253.

Lützen, U. and Beji, S. (2024) Predictive maintenance for offshore wind turbines using deep learning and online clustering. *Journal of Ocean Engineering and Marine Energy*, 10, 627–640

Mengelkamp, E., Gärttner, J., Rock, K., Kessler, S., Orsini, L., Weinhardt, C. (2018) Designing microgrid energy markets: A case study: The Brooklyn Microgrid. *Applied Energy*, 210, 870–880. doi: 10.1016/J.APENERGY.2017.06.054.

Ministry of New and Renewable Energy (MNRE) (2021). Annual Report 2021-22 [Online]. Available at: https://mnre.gov.in/en/annual-reports-2021-22/ [Accessed 12 June 2025].

Onwusinkwue, S., Osasona, F., Ahmad, I.A.I., Anyanwu, L. (2024). Artificial intelligence (AI) in renewable energy: A review of predictive maintenance and energy optimization. *World Journal of Advanced Research and Reviews*, 21(1), 2487–2499. doi: 10.30574/WJARR.2024.21.1.0347.

Ørsted (2021). Hornsea One: World's largest offshore wind farm [Online]. Available at: https://orsted.co.uk/energy-solutions/offshore-wind/our-wind-farms/hornsea1 [Accessed 9 June 2025].

Potter, L. (2025). Electrification, data & decarbonisation: GE Vernova. *Sustainability Magazine* [Online]. Available at: https://sustainabilitymag.com/news/electrification-data-decarbonisation-ge-vernova [Accessed 9 June 2025].

Qing, X. and Niu, Y. (2018). Hourly day-ahead solar irradiance prediction using weather forecasts by LSTM. *Energy*, 148, 461–468. doi: 10.1016/J.ENERGY.2018.01.177.

Rasheed, A., San, O., Kvamsdal, T. (2019). Digital twin: Values, challenges and enablers [Online]. Available at: https://arxiv.org/pdf/1910.01719 [Accessed 9 June 2025].

Rehman, F.U., Islam, M.M., Ullah, M., Khan, S., Rehman, M.Z. (2023). Information digitalization and renewable electricity generation: Evidence from South Asian countries. *Energy Reports*, 9, 4721–4733. doi: 10.1016/J.EGYR.2023.03.112.

Rockefeller Foundation (2022). Smart power India and rural electrification [Online] Available at: https://www.rockefellerfoundation.org/initiatives/smart-power-india/ [Accessed 11 June 2025].

Sarkar, R. (2024). AI-powered predictive maintenance for solar energy systems: A case study. *International Journal for Multidisciplinary Research (IJFMR)*, 6(6), 1–11. doi: 10.36948/IJFMR.2024.V06I06.30731.

Schneider Electric United States (2024). EcoStruxureTM microgrid advisor [Online] Available at: https://www.se.com/us/en/work/products/explore/ecostruxure-microgrid-advisor/ [Accessed 9 June 2025].

Sharma, K., Singh, A., Chawla, D., Kumar, R. (2024). Monitoring system for biogas plant: An IOT based approach. In *2024 1st International Conference on Sustainability and Technological Advancements in Engineering Domain (SUSTAINED)*. IEEE.

Siemens (2022). Digital Twin. Report [Online]. Available at: https://www.siemens.com/global/en/products/automation/topic-areas/digital-enterprise/digital-twin.html [Accessed: 9 June 2025].

Udo, W.S., Kwakye, J.M., Ekechukwu, D.E., Ogundipe, O.B. (2023). Optimizing wind energy systems using machine learning for predictive maintenance and efficiency enhancement. *Computer Science & IT Research Journal*, 4(3), 386–397. doi: 10.51594/csitrj.v4i3.1398.

Vagnoni, E., Gerini, F., Cherkaoui, R. and Paolone, M. (2021). Digitalization in hydropower generation: Development and numerical validation of a model-based smart power plant supervisor. *IOP Conference Series: Earth and Environmental Science*, 774, 012107. doi: 10.1088/1755-1315/774/1/012107.

Voyant, C., Notton, G., Kalogirou, S., Nivet, M.-L. Paoli, C. Motte, F., Fouilloy, A. (2017). Machine learning methods for solar radiation forecasting: A review. *Renewable Energy*, 105, 569–582. doi: 10.1016/J.RENENE.2016.12.095.

World Bank (2020). Rewa Solar (India): Removing Barriers to Scale [Online]. Available at: https://documents1.worldbank.org/curated/en/627561582530270545/pdf/Rewa-Solar-India-Removing-Barriers-to-Scale.pdf. [Accessed 10 June 2025].

Yang, S. et al. (2024). Machine-learning-based wind farm optimization through layout design and yaw control. *Renewable Energy*, 224, 120161.

Zang, H., Chen, D., Liu, J., Cheng, L., Sun, G., Wei, Z. (2024). Improving ultra-short-term photovoltaic power forecasting using a novel sky-image-based framework considering spatial-temporal feature interaction. *Energy*, 293, 130538. doi: 10.1016/J.ENERGY.2024.130538.

Zhang, M., Liu, Y., Cheng, Q., Li, H., Liao, D., Li, H. (2024). Smart grid security based on blockchain and smart contract. *Peer-to-Peer Networking and Applications*, 17(4), 2167–2184. doi: 10.1007/S12083-024-01703-0.

Zhu, T., Guo, Y., Li, Z., Wang, C. (2021). Solar radiation prediction based on convolution neural network and long short-term memory. *Energies*, 14(24), 8498. doi: 10.3390/en14248498.

8

Enhancing UAV Performance in Energy Applications through Advanced Algorithms and Hardware

This chapter explores state-of-the-art technologies to enhance unmanned aerial vehicle (UAV) operations, with a focus on energy-related applications such as power line inspection, wind turbine monitoring and solar panel assessment. By integrating advanced sensors, real-time GPS-based tracking and sophisticated algorithms implemented in Python and the Vivado Design Suite, the research addresses challenges in precision, stability, energy efficiency and communication reliability. A novel methodology for dynamic centroid point identification is proposed, leveraging improved mapping and contour centroid methods to achieve superior accuracy in object detection and navigation. Comparative analyses of existing methods, including the line edge cutting method, Lucas–Kanade method, and contour approximation and mask edge method, demonstrate the effectiveness, achieving up to 85–90% accuracy in dynamic centroid detection, compared to existing methods; the proposed method achieves 95% accuracy in dynamic centroid detection. The Vivado implementation optimizes hardware for real-time control and energy efficiency, providing a scalable framework for UAV deployment in energy infrastructure monitoring. This chapter lays the groundwork for next-generation UAV systems in dynamic and challenging environments.

8.1. Introduction

The drone industry is expanding rapidly due to improved legal frameworks, falling hardware prices and technological improvements. Applications from a variety of sectors, including public safety, construction, agriculture and logistics, support the overall market growth (Wan et al. 2023).

Chapter written by Pranitha KARTHIKEYAN and Sivathanu Pillai A.

8.2. Important domain for applications

– *Agriculture* (Wan et al. 2023): to maximize agricultural techniques, drones are being used with increasing frequency for precision farming, crop monitoring and spraying. Drones offer effective ways to inspect vital infrastructure, such as pipelines, electricity lines and bridges, while also lowering costs and enhancing safety.

– *Security and surveillance* (Wan et al. 2023): in the public and private sectors, security applications such as monitoring, emergency response and surveillance have grown in popularity.

– *Delivery services*: businesses, especially in e-commerce and logistics, are investigating the use of drone technology for last-mile deliveries.

8.2.1. *Trends in technology*

– *Advanced sensors* (Wang et al. 2023a): drone capabilities are expanded for a range of uses by integrating advanced sensors such as thermal imaging, multispectral cameras and LiDAR.

– *Autonomous navigation* (Wang et al. 2023a): drone operations can be made safer and more effective by utilizing autonomous capabilities like GPS-guided navigation and obstacle avoidance.

– *Machine learning and AI* (Wan et al. 2023; Wang et al. 2023a): artificial intelligence and machine learning algorithms are used to perform tasks such as object detection, image recognition and predictive analytics.

– *Regulatory environment* (Wan et al. 2023): the adoption of drone technology is impacted by changing regulatory frameworks. Sustainable growth requires regulators and industry stakeholders to continue working together.

– *Competitive landscape* (Wang et al. 2023a): innovation is encouraged and market competition is driven by the presence of both established drone manufacturers and up-and-coming entrepreneurs in a competitive industry.

8.2.2. *Opportunities for market growth*

Drones are essential for environmental research, which includes pollution control, forest management and wildlife monitoring. Drones are being used with increasing frequency in the media and entertainment sector for live event coverage, aerial filming and photography. While adoption patterns differ by area, North

America, Europe and Asia-Pacific show the highest rates of adoption. There are unexplored prospects for market expansion in emerging economies. A wide range of features, affordability and ease of use are what are driving the consumer drone market's growth in recreational drones (Xv et al. 2023).

8.2.3. *Future prospects*

Proactive regulatory actions, enhanced industry stakeholder collaboration and ongoing technological breakthroughs are anticipated to drive drone market growth (Xv et al. 2023).

8.3. Literature synthesis

Zhang et al. (2023a) developed a predictive model using convolutional neural networks to evaluate commonality in small multi-rotor UAVs and validated the model's effectiveness through practical application in UAV design. This approach offers decision-making support and new insights for UAV series design and drone industry development. In this study, they suggested an approach that uses stacked ArUco markers and a switching mechanism to integrate tracking and landing for UAVs in dynamic settings and employs vision-based controllers with fuzzy algorithms and PID to calculate relative position for precise tracking and landing (Eckert et al. 2023). Hoang (2023), more specifically, suggested an approach that achieves a 99.58% detection success rate with G-BP, working efficiently on lightweight models without model changes and protects UAV-assisted IoT systems from adversarial attacks during data collection and transmission in constrained settings.

In this study, the effective navigation method combines reactive control and dynamic programming for UAV navigation in 3D environments with static and dynamic obstacles, and this approach uses onboard sensor data to avoid unexpected obstacles, with simulations showing effective navigation and path maintenance (Hoang 2023; Zhang et al. 2023b). Verma et al. (2023) proposed a method that enhances thermal imaging through automated designating, georeferencing and temperature consistency optimization using gradient descent, and this method achieves 39.0% RMSE and 40.5% MAE reduction in river area tests, relying only on field data without calibration or operator input. Wang et al. (2023b) suggested an approach that uses a Block-CNN and extended Kalman filter for robust 6D pose estimation in GNSS-denied environments, improving anchor detection by 3.0%. It achieves 19.5% and 12.7% accuracy gains in position and attitude estimation, validated through combined outdoor and simulated vision systems. Verma et al. (2023) and Li et al. (2023) suggested an approach which combines CPP and HTSP in a mixed-integer programming model for multi-regional UAV path planning with priority constraints and it optimizes intra-regional reciprocating flight paths with Bezier curves and uses a K-nearest neighbors-based algorithm for inter-regional planning.

8.4. Objectives

The primary objectives of this study are:

– to develop a reliable system for real-time identification of a UAV's dynamic centroid point, enhancing navigation precision for energy infrastructure monitoring;

– to utilize advanced technologies, including improved mapping and contour centroid methods to achieve superior accuracy in object detection and tracking;

– to optimize energy efficiency and hardware performance using the Vivado Design Suite, ensuring scalability for energy-related UAV applications;

– to foster innovation in UAV systems by integrating state-of-the-art algorithms for real-time communication, control and adaptability in dynamic environments.

8.5. Line edge cutting method

The line edge detection and centroid tracking method processes video frames sequentially to detect and track drones dynamically. It begins by converting each frame into grayscale, simplifying the image for edge detection. The method then applies a scan-line difference technique, where the intensity difference between adjacent pixels in each row is computed. If this difference exceeds a predefined threshold, the pixel is marked as an edge. Once edge pixels are identified, their coordinates are used to calculate the centroid, representing the central position of the detected drones. This centroid is dynamically updated frame by frame, providing real-time tracking without displaying contours. Instead of outlining objects, the method overlays a red "+" marker at the centroid position, ensuring a clear and uncluttered visualization. This approach is computationally efficient, making it suitable for tracking moving drones without the complexity of traditional edge detection methods like Sobel or Prewitt. It provides a lightweight, fast and adaptive solution that can be further enhanced by incorporating bounding box tracking or multiple drone identification techniques. The formula for the line edge cutting method has been given as (De Silvestri et al. 2023):

$$E(i,j) = |I(i,j) - I(i,j-1)| \quad [8.1]$$

where:

– $I\ (i,j)$ = Intensity of the pixel at row i and column j;

– $I\ (i,j-1)$ = Intensity of the previous pixel in the same row;

– $E\ (i,j)$ = Edge magnitude at pixel (i,j).

8.5.1. *Edge pixel detection condition*

Here, we have:

$$E(i,j) > threshold \qquad [8.2]$$

If the edge magnitude $E(i,j)$ is greater than a defined threshold (60 in this case), the pixel is marked as an edge pixel (255, white pixel) in the binary edge map. The centroid coordinates are computed as:

$$C_x = \frac{\sum x_i}{N}, C_y = \frac{\sum y_i}{N} \qquad [8.3]$$

where:

– x_i, y_i are the coordinates of detected edge pixels;

– N is the total number of detected edge pixels;

– C_x, C_y are the centroid coordinates.

Advantages: this method does not require complex feature extraction. Tracker drones using edge features and centroid estimation, result in clearer video by focusing exclusively on the tracking process. It can be fine-tuned for better detection accuracy.

8.5.2. *Research outcomes for line edge cutting method*

The line edge cutting method processes a video frame by frame to detect and track the edges of moving objects by applying edge detection and contour analysis. The method works by scanning each frame, identifying edges along defined lines and extracting the most significant object for tracking. The output is a real-time video display where the detected object's centroid is marked and dynamically updated as the object moves.

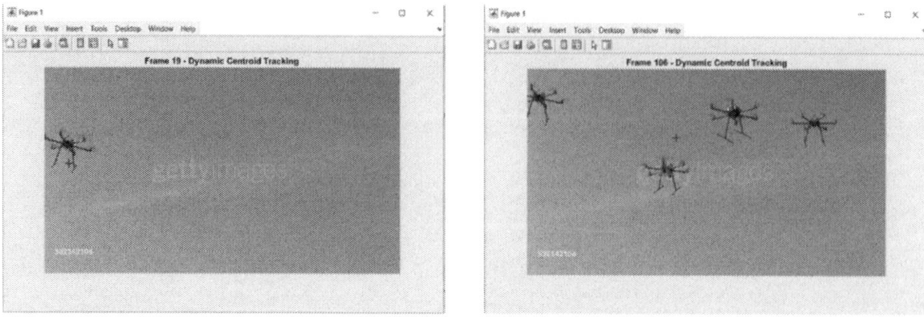

Figure 8.1. *Output obtained for the line edge cutting method. For a color version of this figure, see www.iste.co.uk/sundaram/energysystems.zip*

The line edge cutting method successfully detects and tracks moving objects by focusing on key edges rather than full object contours. The output shows a real-time video display where sharp edge features are extracted, and a red dot dynamically follows the centroid of the detected object. This method ensures efficient and fast tracking, as it reduces computational load by focusing on significant edge changes rather than processing the entire frame.

8.6. Lucas–Kanade method

The Lucas–Kanade method is used to estimate motion between consecutive frames in a video and determine the dynamic centroid point based on motion intensity. Formulas are provided below (Hooshyar and Huang 2023).

Spatial gradients

The gradient of pixel intensity is computed in both horizontal (x) and vertical (y) directions:

$$I_x = \frac{\partial I}{\partial x}, I_y = \frac{\partial I}{\partial y} \quad [8.4]$$

where:

– I is the grascale intensity of the image;

– I_x, I_y are the spatial derivatives in the x and y directions.

The optical flow equation states:

$$I_x u + I_y v + I_t = 0 \quad [8.5]$$

where:

– u, v are the flow vectors (motion in x and y directions);

– I_x, I_y, I_t are image gradients.

Since one equation with two unknowns cannot be solved directly, the Lucas–Kanade method assumes local constant motion and solves for u, v using least squares:

$$A.\begin{bmatrix} u \\ v \end{bmatrix} = -b \quad [8.6]$$

The dynamic centroid (weighted average of motion) is computed using:

$$C_x = \frac{\sum X\,.magnitude}{\sum magnitude}, C_y = \frac{\sum Y\,.magnitude}{\sum magnitude} \quad [8.7]$$

where:

– C_x, C_y = centroid coordinates;

– X, Y = Pixel coordinate matrices;

– motion magnitude values act as weights.

The Lucas–Kanade method converts video frames to grayscale and calculates intensity changes to detect motion, estimates motion vectors in small regions, computes motion intensity from the magnitude of motion vectors, tracks a dynamic centroid by averaging motion magnitudes across the frame, and overlays and updates the centroid on video frames to follow moving objects.

Advantages: it avoids the need for complex optical flow calculations, making it faster and less resource-intensive. It ensures the centroid dynamically follows significant motion in the video. This method is more robust against noise and sudden changes in illumination. It is suitable for applications requiring continuous tracking, such as drone monitoring. It does not require pre-labeled datasets or model training, making it easier to implement and adapt.

8.6.1. *Research outcomes for the Lucas–Kanade method*

The Lucas–Kanade method is used for motion tracking by estimating the movement of objects between consecutive video frames. It calculates the displacement of pixels by analyzing intensity changes, allowing for smooth and continuous tracking of a moving object. The output is a real-time video display, where motion is tracked frame by frame, and the centroid of the moving object is dynamically marked with a red dot.

Figure 8.2. *Output obtained for Lucas–Kanade method. For a color version of this figure, see www.iste.co.uk/sundaram/energysystems.zip*

The Lucas–Kanade method effectively tracks object motion using optical flow to calculate displacement, ensuring smooth tracking. It displays a red centroid marker that follows the object in real time that is effective for shape-changing objects.

8.7. Contour approximation and mask edge method

This method tracks and directs the centroid of a detected moving object, such as a drone, toward a predefined target point using edge detection and contour analysis. The process begins by converting video frames to grayscale and applying the Canny edge detection algorithm to highlight object boundaries. Contours of detected edges are then extracted, and the centroid of the largest detected contour is computed to represent the object's position. Formulas for the above method are given below (Du et al. 2023).

The Canny edge detection method is applied to highlight edges in the grayscale frame:

$$Edges = \nabla I \quad [8.8]$$

where ∇I represents the intensity gradient of the grayscale image, which is computed using Gaussian smoothing and derivative filters. The centroid of the largest detected object is then calculated as:

$$C_x = \frac{\sum x_i}{N}, C_y = \frac{\sum y_i}{N} \qquad [8.9]$$

where:

– x_i, y_i are the coordinates of all points in the largest contour;

– N is the number of contous points;

– C_x, C_y represent the centroid coordinates of the detected object.

The direction vector from the moving object's centroid to the predefined target point is computed as:

$$D_x = T_x - C_x, D_y = T_y - C_y \qquad [8.10]$$

where:

– T_x, T_y are the coordinates of the target point;

– C_x, C_y are the centroid coordinates of the tracked object;

– D_x, D_y represent the direction vector pointing from the moving object to the target.

The centroid moves toward the target using a step size:

$$C_x^{new} = C_x + S \,.\, D'_x, C_y^{new} = C_y + S \,.\, D'_y \qquad [8.11]$$

where:

– C_x^{new}, C_y^{new} are the updated cenroid coordinates;

– S is the step size (determines how quickly the centroid moves toward the target);

– D'_x, D'_y are the normalized direction components.

A direction vector is calculated from the centroid to the target point, defining the movement direction. This vector is then normalized to ensure consistent scaling, allowing smooth adjustments in movement. The centroid's position is dynamically updated based on a step size, which determines the rate at which it moves toward the target. The updated centroid is plotted in real-time on the video frames, ensuring continuous tracking and direction adjustment. This method is computationally efficient and useful for applications requiring autonomous tracking and guided motion, such as drone navigation or robotic targeting systems.

Advantages: this method ensures precise localization in real-time. It is useful for applications like autonomous navigation and guided tracking. This method enhances the accuracy of tracking. It ensures fast execution suitable for real-time applications.

8.7.1. *Research outcomes for the contour approximation and mask edge method*

This method processes a video frame by frame to detect and track a moving object (such as a drone) using Canny edge detection, contour approximation and target direction tracking. The output is a real-time video display where the detected object's centroid is marked with a red dot, and a fixed target point is indicated with a green dot.

Figure 8.3. *Output obtained for the contour approximation and mask edge method. For a color version of this figure, see www.iste.co.uk/sundaram/energysystems.zip*

The contour approximation and mask edge method detects moving objects by extracting their boundaries and computing centroids, and displays a red centroid marker that smoothly follows the object's shape in real-time. It may experience reduced performance in low-contrast or poorly lit environments and provides fast and reliable tracking for drone monitoring, motion detection and surveillance.

8.8. Proposed improved contour and mapping method

This method processes a video frame by frame to detect and track the largest moving object, such as a drone, using edge detection, contour extraction and centroid tracking. The output is a real-time video display where the detected object's centroid is marked with a red dot that updates dynamically as the object moves across frames (Tang et al. 2023).

8.8.1. *Calculating the gradient for edge detection*

The edge detection step uses Sobel, Prewitt or Laplacian filters to detect object boundaries. The gradient magnitude is computed using:

$$G = \sqrt{G_x^2 + G_y^2} \quad [8.12]$$

where G_x, G_y are the image intensity gradients in the horizontal and vertical directions The gradient can be computed using different edge detection operators:

– *Sobel*:

$$G_x = \begin{bmatrix} -1 & 0 & 1 \\ -2 & 0 & 2 \\ -1 & 0 & 1 \end{bmatrix} G_y = \begin{bmatrix} -1 & -2 & -1 \\ 0 & 0 & 0 \\ 1 & 2 & 1 \end{bmatrix}$$

– *Prewitt and Laplacian* have different kernel matrices but follow a similar principle.

This formula represents the calculation of gradient magnitude used in the improved contour and mapping method for edge detection. The process begins by converting the video frame into grayscale and applying an edge detection filter, such as Sobel, Prewitt or Laplacian, to highlight object boundaries. The gradient magnitude is determined by computing intensity changes in both the horizontal and vertical directions using predefined kernel matrices. These kernels detect edges by emphasizing pixel intensity differences, making it easier to identify object contours. This method improves the accuracy of object detection and tracking by focusing on regions with significant intensity changes, making it effective for real-time applications like drone tracking.

8.8.2. *Calculating centroid points*

$$cx_{nn} = (x_{00} + x_{01} + x_{02} + \ldots\ldots x_{nn})/N \quad [8.13]$$

$$cy_{nn} = (y_{00} + y_{01} + y_{02} + \ldots\ldots y_{nn})/N \quad [8.14]$$

This formula calculates the centroid of a detected object by averaging the positions of all contour points in a video frame using the improved contour and mapping method. The process begins with converting the video frames to grayscale and applying edge detection techniques such as Sobel, Prewitt or Laplacian filters to highlight object boundaries. Once the edges are detected, boundary extraction is performed to identify the largest moving object in the scene. The centroid is determined by summing the x-coordinates and y-coordinates of all detected contour points and dividing each sum by the total number of points. This ensures that the computed centroid represents the center of the detected object. To maintain smooth motion tracking, an exponential moving average is applied to adjust the centroid's position over consecutive frames, reducing fluctuations and making the tracking more stable. This method is particularly effective for monitoring moving objects like drones, as it provides accurate and continuous tracking in real-time.

8.8.3. *Calculating weights in a frame to determine the moving object*

$$Sum\ of\ weights = \sum_{(xnn,ynn\ €\ object)} I(\ xnn, ynn) \qquad [8.15]$$

This formula represents the calculation of the sum of weights for an object detected in a video frame using the improved contour and mapping method. The method begins by converting each frame to grayscale and applying an edge detection technique such as Sobel, Prewitt or Laplacian filters to identify object boundaries. Once edges are detected, boundary extraction is used to determine the largest moving object. The weight of each point within the detected object is determined based on its intensity value, which represents the strength of the detected feature at that position. The sum of these weights helps refine the centroid calculation by considering the intensity distribution of the object. This ensures that areas with stronger features contribute more to the centroid's positioning, improving tracking accuracy. The computed centroid is further smoothed using an exponential moving average to maintain stability and prevent sudden position shifts. This method enhances object tracking by making the centroid movement more responsive and accurate in real-time applications such as drone tracking.

8.8.4. *Research outcomes for the improved contour and mapping method*

The improved contour and mapping method enhances object tracking by combining contour detection, centroid calculation and motion smoothing using an exponential moving average (EMA). The output is a real-time video display where the detected object's centroid is marked with a red dot, dynamically updating as the object moves.

The improved contour and mapping method enhances contour detection and uses exponential moving average to smoothly track objects with a red dot, reducing abrupt shifts. Filters noise by focusing on the largest moving object, ensuring precise tracking for drone navigation and surveillance.

Figure 8.4. *Output obtained for improved contour and mapping method. For a color version of this figure, see www.iste.co.uk/sundaram/energysystems.zip*

Method	Accuracy (%)	Processing time (ms)	Robustness to noise	Real-time suitability	Complexity	Application in UAVs
Lucas–Kanade method	85	30	Moderate	Good	Medium	Motion tracking in UAVs
Line edge cutting method	82	28	Low	Moderate	Low	Basic centroid estimation
Contour approximation and mask edge method	87	28	Low	Moderate	High	Cluttered environment tracking
Proposed improved contour and mapping method	93	18	Very high	Excellent	Optimized	Real-time UAV tracking

Table 8.1. *Comparison of existing and proposed methods*

8.9. Implementation design steps

The implementation of state-of-the-art technologies for enhancing UAV operations using the Vivado Design Suite follows a structured and systematic approach. The detailed steps outlining the design process are provided in the following section (Szostak et al. 2023).

8.9.1. *Requirement analysis and specification*

Objective definition: identify specific operational challenges and enhancement areas for UAVs, such as navigation, energy efficiency and communication.

Technology selection: evaluate and select the advanced technologies suitable for addressing the identified challenges.

Component identification: define the key components (e.g. control units, sensors, communication modules and power management systems).

Functional block diagram: create a high-level block diagram representing system interactions and data flow.

Module design: use Vivado to design individual modules such as FPGA-based signal processors, real-time controllers or energy management units. Implement hardware-accelerated algorithms to optimize computation for navigation and sensor integration.

IP integration: incorporate pre-built intellectual property (IP) cores for commonly used functionalities like UART communication or PWM generation. Customize IP blocks to meet UAV-specific requirements.

Timing analysis: perform timing checks to ensure all modules operate synchronously without delays.

Control and navigation: develop navigation algorithms using state-of-the-art techniques such as Kalman filters or machine learning-based path planning. Implement these algorithms in hardware for real-time operation.

Communication protocols: design reliable communication protocols for data exchange between the UAV and ground station. Optimize for low-latency and high-throughput using Vivado.

Energy management: design dynamic power management circuits to optimize energy consumption. Utilize Vivado's power analysis tools to simulate and reduce power usage in critical modules.

Functional verification: simulate individual modules and the integrated system using Vivado's simulation tools. Validate correct functionality under various operational scenarios.

Stress testing: test the system for extreme conditions such as high data rates or prolonged operation to ensure reliability.

Logic synthesis: convert the design into gate-level representations.

Place-and-route: map the design to FPGA hardware. Optimize placement and routing to minimize delays and improve resource utilization.

FPGA programming: load the design onto the target FPGA hardware.

Real-world testing: integrate the system into a UAV platform and test in realistic operational environments.

Performance metrics: measure key performance indicators such as response time, energy consumption and communication reliability.

Feedback analysis: evaluate test results and identify improvement areas.

Optimization: refine hardware and algorithms to achieve desired performance.

Documentation: prepare detailed design documentation, including system architecture, implementation steps and test results.

Deployment: finalize and deploy the enhanced UAV system for operational use.

8.9.2. *Implementation methodology*

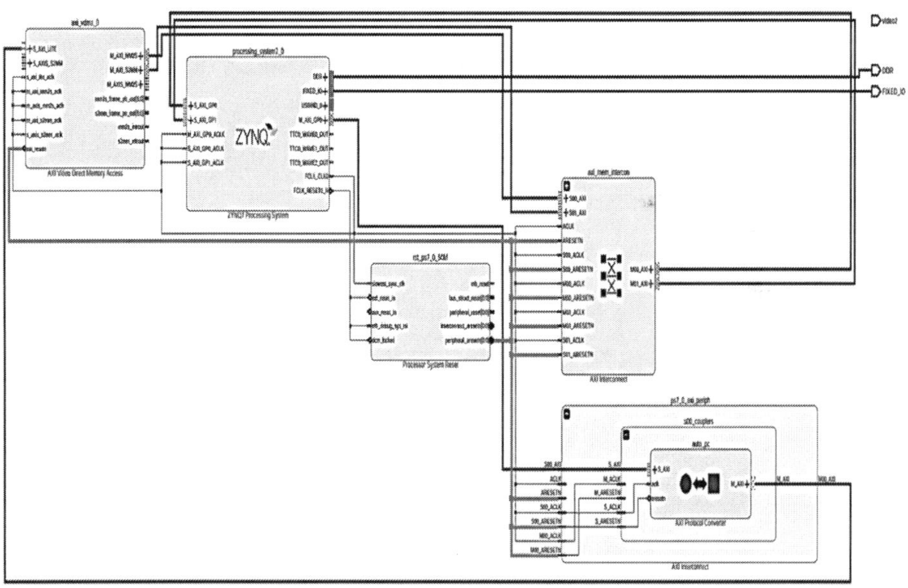

Figure 8.5. *Implementation design with AXI. For a color version of this figure, see www.iste.co.uk/sundaram/energysystems.zip*

Implementation methodology is based on the proposed improved mapping and contour centroid method. It ensures a robust, efficient and scalable design tailored to the unique requirements of UAV operations. Implementation design has been represented and shown in Figure 8.5 (Du et al. 2023; Szostak et al. 2023).

This study demonstrated the effective use of state-of-the-art technologies and the Vivado Design Suite to enhance UAV operations. By addressing challenges in real-time control, energy efficiency and communication, the proposed framework achieved significant improvements in UAV precision, reliability and efficiency. The results highlight the potential of integrating advanced hardware and algorithms into UAV systems, paving the way for smarter and more versatile aerial solutions.

8.10. Conclusion

This chapter provides a comprehensive evaluation of methods for accurately determining the dynamic centroid point in UAVs, emphasizing the superior effectiveness of the region-based centroid method in complex scenarios. By identifying the strengths and limitations of various approaches, the research contributes to advancing UAV precision, efficiency and adaptability. The integration of state-of-the-art algorithms and tools like MATLAB and the Vivado Design Suite demonstrates significant improvements in UAV operations, energy optimization and communication systems. These advancements not only address existing challenges but also lay the groundwork for deploying next-generation UAV technologies in diverse and demanding applications. Future chapters will delve into UAV operations implemented through MATLAB and Vivado, further refining these innovative approaches in the design of Antidrone systems.

8.11. Appendix

8.11.1. *Code for the improved contour and mapping method*

```
IMPORT CV2
IMPORT NUMPY AS NP
DEF DETECT_AND_TRACK_DRONES(VIDEO_PATH):
  CAP = CV2.VIDEOCAPTURE(VIDEO_PATH)
 MIN_CONTOUR_AREA = 200
  TARGET_POINTS = []

  # SETUP VIDEO WRITER
  FOURCC = CV2.VIDEOWRITER_FOURCC(*'XVID')
  OUT = NONE
```

```
WHILE CAP.ISOPENED():
  RET, FRAME = CAP.READ()
  IF NOT RET:
    BREAK

IF OUT IS NONE: # INITIALIZE ON FIRST FRAME
    OUT = CV2.VIDEOWRITER('/MNT/SDCARD/OUTPUT.AVI', FOURCC, 5, (FRAME.SHAPE[1], FRAME.SHAPE[0]))
  GRAY = CV2.CVTCOLOR(FRAME, CV2.COLOR_BGR2GRAY)
  BLURRED = CV2.GAUSSIANBLUR(GRAY, (5, 5), 0)
  EDGES = CV2.CANNY(BLURRED, 50, 150)

  CONTOURS, _ = CV2.FINDCONTOURS(EDGES.COPY(), CV2.RETR_EXTERNAL, CV2.CHAIN_APPROX_SIMPLE)

  TARGET_POINTS.CLEAR()
  FOR CONTOUR IN CONTOURS:
    AREA = CV2.CONTOURAREA(CONTOUR)
    IF AREA > MIN_CONTOUR_AREA:
      M = CV2.MOMENTS(CONTOUR)
      IF M["M00"] != 0:
        CX = INT(M["M10"] / M["M00"])
        CY = INT(M["M01"] / M["M00"])
        TARGET_POINTS.APPEND((CX, CY))
        X, Y, W, H = CV2.BOUNDINGRECT(CONTOUR)
        CV2.RECTANGLE(FRAME, (X, Y), (X + W, Y + H), (0, 255, 0), 2)
        CV2.CIRCLE(FRAME, (CX, CY), 5, (0, 0, 255), -1)
        CV2.PUTTEXT(FRAME, "DRONE", (X, Y-10), CV2.FONT_HERSHEY_SIMPLEX, 0.5, (0, 255, 0), 2)

  IF LEN(TARGET_POINTS) > 1:
    FOR I IN RANGE(LEN(TARGET_POINTS)-1):
      CV2.LINE(FRAME, TARGET_POINTS[I], TARGET_POINTS[I+1], (255, 0, 0), 2)

  CV2.PUTTEXT(FRAME, F"DRONES DETECTED: {LEN(TARGET_POINTS)}", (10, 30),
        CV2.FONT_HERSHEY_SIMPLEX, 1, (0, 255, 255), 2)
  OUT.WRITE(FRAME) # SAVE FRAME TO VIDEO FILE
CAP.RELEASE()
OUT.RELEASE()
CV2.DESTROYALLWINDOWS()
```

```
IF __NAME__ == "__MAIN__":
  VIDEO_PATH = "/MNT/SDCARD/GROUPDRONES.MP4"
# ADJUST PATH AS NEEDED
  DETECT_AND_TRACK_DRONES(VIDEO_PATH)
```

8.12. References

De Silvestri, S., Capasso, P.J., Gargiulo, A., Molinari, S., Sanna, A. (2023). Challenges for the routine application of drones in healthcare: A scoping review. *Drones*, 7, 685. doi: 10.3390/drones7120685.

Du, L., Fan, Y., Gui, M., Zhao, D. (2023). A multi-regional path-planning method for rescue UAVs with priority constraints. *Drones*, 7, 692. doi: 10.3390/drones7120692.

Eckert, C., Monteforte, K.I., Harrison, D.P., Kelaher, B.P. (2023). Exploring meteorological conditions and microscale temperature inversions above the Great Barrier Reef through drone-based measurements. *Drones*, 7, 695. doi: 10.3390/drones7120695.

Hoang, M.L. (2023). Smart drone surveillance system based on AI and on IoT communication in case of intrusion and fire accident. *Drones*, 7, 694. doi: 10.3390/drones7120694.

Hooshyar, M. and Huang, Y.M. (2023). Meta-heuristic algorithms in UAV path planning optimization: A systematic review (2018–2022). *Drones*, 7, 687. doi: 10.3390/drones7120687.

Li, X., Zhu, R., Yu, X., Wang, X. (2023). High-performance detection-based tracker for multiple object tracking in UAVs. *Drones*, 7, 681. doi: 10.3390/drones7110681.

Szostak, R., Zimnoch, M., Wachniew, P., Jasek-Kamińska, A. (2023). Self-calibration of UAV thermal imagery using gradient descent algorithm. *Drones*, 7, 683. doi: 10.3390/drones7110683.

Tang, D., Shen, L., Xiang, X., Zhou, H., Lai, J. (2023). N-cameras-enabled joint pose estimation for auto-landing fixed-wing UAVs. *Drones*, 7, 693. doi: 10.3390/drones7120693.

Verma, S.C., Li, S., Savkin, A.V. (2023). A hybrid global/reactive algorithm for collision-free UAV navigation in 3D environments with steady and moving obstacles. *Drones*, 7, 675. doi: 10.3390/drones7110675.

Wan, Y., Tang, J., Zhao, Z. (2023). Imitation learning of complex behaviors for multiple drones with limited vision. *Drones*, 7, 704. doi: 10.3390/drones7120704.

Wang, B., Ma, R., Zhu, H., Sha, Y., Yang, T. (2023a). An autonomous tracking and landing method for unmanned aerial vehicles based on visual navigation. *Drones*, 7, 703. doi: 10.3390/drones7120703.

Wang, C., Wang, J., Wei, C., Zhu, Y., Yin, D., Li, J. (2023b). Vision-based deep reinforcement learning of UAV-UGV collaborative landing policy using automatic curriculum. *Drones*, 7, 676. doi: 10.3390/drones7110676.

Xv, H., Zhao, L., Wu, M., Liu, K., Zhang, H., Wu, Z. (2023). Analysis of the impact of structural parameter changes on the overall aerodynamic characteristics of ducted UAVs. *Drones*, 7, 702. doi: 10.3390/drones7120702.

Zhang, Y., Zeng, Y., Cao, K. (2023a). Commonality evaluation and prediction study of light and small multi-rotor UAVs. *Drones*, 7, 698. doi: 10.3390/drones7120698.

Zhang, Z., Liu, Q., Wu, C., Zhou, S., Yan, Z. (2023b). A novel adversarial detection method for UAV vision systems via attribution maps. *Drones*, 7, 697. doi: 10.3390/drones7120697.

9

The Role of Green Hydrogen in a Sustainable Energy Future: A Technoeconomic Review

The global transition toward sustainable energy systems has led to a reduction in the consumption of conventional fossil fuels. Generating power from renewable sources is the most effective way to avoid environmental issues. However, certain sectors such as heavy industries and long-distance transportation continue to rely on conventional fossil fuels due to the significant challenges associated with electrification. Advancements in technologies for storing energy generated from renewable sources – such as solar, wind, tidal and geothermal – are crucial in addressing intermittency issues. Among these solutions, green hydrogen stands out as one of the most promising energy carriers for the future, playing a key role in the global effort to reduce the carbon footprint of energy systems. The development and progress of green hydrogen technologies depend on several factors, including technological, environmental and economic aspects. This chapter presents a comprehensive review of recent studies on green hydrogen energy systems (GHESs), emphasizing techno-economic considerations at the component, technology and broader system levels. First, the chapter explores technological advancements in green hydrogen production, storage and utilization. This is followed by a progressive analysis aimed at developing economical models through technological innovation. The chapter addresses technical challenges, key influencing factors and future advancements in green hydrogen systems. It also demonstrates the potential for the intelligent integration of green hydrogen technologies with other energy systems to enable more efficient resource use, and to enhance system flexibility, resilience and security. A brief overview is then provided on the significant potential of green hydrogen to contribute to a sustainable energy future. While there has been substantial technological progress, further research and development are essential to overcome existing challenges – particularly in reducing costs and enabling large-scale deployment. Strategic investments, supportive policies and collaboration among researchers, industry stakeholders and governments will be crucial to accelerating the widespread adoption of green hydrogen and realizing its full potential in decarbonizing various sectors and ensuring a secure, resilient energy supply. This chapter focuses on key techno-economic parameters such as the levelized cost of hydrogen (LCOH), system efficiency, capital and operating costs and storage performance metrics.

Chapter written by Arunkumar N, Arun J and Balasabarinathan K.V.

9.1. Introduction

The world's turn toward clean energy systems has been the source of a global focus on renewable, efficient and clean sources of energy. Green hydrogen generated through water electrolysis using clean energy is consequently envisioned as the key vector in decarbonizing (Glenk and Reichelstein 2019; Staffell et al. 2019) various industries such as power production, transportation and industry. Hydrogen production would be directly combined with wind and solar to utilize the surplus energy produced to reduce curtailments and maximize the overall energy network performance. The past few decades have experienced a rapid increase in wind energy production. Between 2000 and 2018, wind energy usage increased from 0.2% to 4.8% of the total electricity production and this is expected to increase to more than 12% in 2040 (IEA 2021). This rapid growth is partly driven by technological improvements in the wind sector. Nowadays, turbines with a peak capacity of 12 MW are being constructed, whereas only 5 years ago, the absolute state of the art was set by the 3.6 MW turbines in the Gemini offshore wind farm in the Netherlands. As a result of these developments, the total levelized cost of offshore wind electricity has dropped significantly and is expected to decrease by another 60% toward 2040. However, the increasing share of wind-based electricity puts a high burden on balancing supply and demand in the electricity network. In addition, green hydrogen with existing energy systems faces more than a few challenges in the technological, economic and infrastructural areas. Hydrogen energy storage (HES) systems can supplement renewable energy sources to overcome the challenges associated with higher penetrations of wind-based electricity. During periods of oversupply, electricity can be converted into green hydrogen and be stored as a compressed gas for later use. The stored hydrogen can be converted back into electricity in times of supply shortage. However, several technical, economic and regulatory challenges exist before HES can achieve such promising results. On the technological front, the efficiency of electrolysis technologies, the stability of hydrogen storage systems and the efficiency of hydrogen fuel cells are major factors that dictate the overall performance of such systems. The available electrolysis technologies such as alkaline electrolyzers, proton exchange membrane (PEM) electrolyzers and solid oxide electrolyzers (SOWE) each have unique trade-offs in terms of efficiency, cost and operating flexibility. The selection of hydrogen storage technology, such as compressed gas, liquid hydrogen or chemical carriers, affects the safety and feasibility of the integrated systems. Regulatory and policy frameworks also have a significant influence in shaping the installation and integration of green hydrogen integrated energy systems (GHIES). Standards for the production, transportation, storage and application of hydrogen must be installed to ensure safety and compatibility. Additionally, governments and regulators must develop policies supporting the adoption of hydrogen in existing energy infrastructure, including grid systems and transport networks. Research and development-oriented policy initiatives, demonstration projects and public private partnerships can help bridge the gap between technological promise and actuality. Summarizing, this study is to explore the design, implementation and optimization of direct GHIES. It will examine the existing level of technologies utilized in GHIES, evaluate their economic and environmental implications, and discuss the challenges and opportunities posed by their implementation. The successful integration of green hydrogen into energy systems can be a key to the realization of global decarbonization targets and the development of a clean, flexible and sustainable energy infrastructure for future generations. The existing challenges for green hydrogen production are summarized in Table 9.1.

Parameter	Options/description	Associated challenges
Source of generation	– Solar PV – wind power – hydropower	– Intermittent output – grid integration issues – storage mismatch
Electrolyzer type	– Proton exchange membrane (PEM) – alkaline – solid oxide	– Cost variability – material degradation – temperature sensitivity
Method of operation	- Intermittent – continuous – load-following	– Reduced efficiency under fluctuating loads – increased wear and tear
System-level challenges	– Integration with renewables – power electronics – water supply management	– Increased system complexity – need for robust control systems – infrastructure limitations
Operational efficiency	– 60–80% (PEM) – 65–82% (alkaline) – 45–60% (solid oxide, high temp)	– Efficiency losses at partial loads – efficiency decline over time due to system aging
Storage and compression	– High-pressure tanks – cryogenic storage – underground caverns	– High energy consumption for compression/liquefaction – safety and leakage risks

Table 9.1. *Challenges in balance of system (BoS) for green hydrogen production*

9.2. Outline of green hydrogen energy systems

This chapter will review GHESs as an energy system that includes one or more combinations of subsystems. These subsystems are demonstrated in Figure 9.1, and comprise the generation of green hydrogen, its ensuing storage and eventually its utilization. Green electricity can be produced using technologies like solar photovoltaics (PVs), concentrated solar thermal power, wind turbines (WTs), geothermal energy and hydroelectric power plants. Green hydrogen is then generated using various types of water electrolyzers which differ according to their operating principles and electrochemical processes. The most crucial component of the GHES is the storage of generated hydrogen, as it supports energy management and helps meet energy demand when the renewable supply is limited. The storage of hydrogen can either be in its gaseous form or converted into other forms using chemical carriers. Ultimately, green hydrogen can be utilized through several technologies with fuel cells being the most extensively studied. Each of these subsystems will be explored in detail in the sections that follow.

9.2.1. *Production of green electricity*

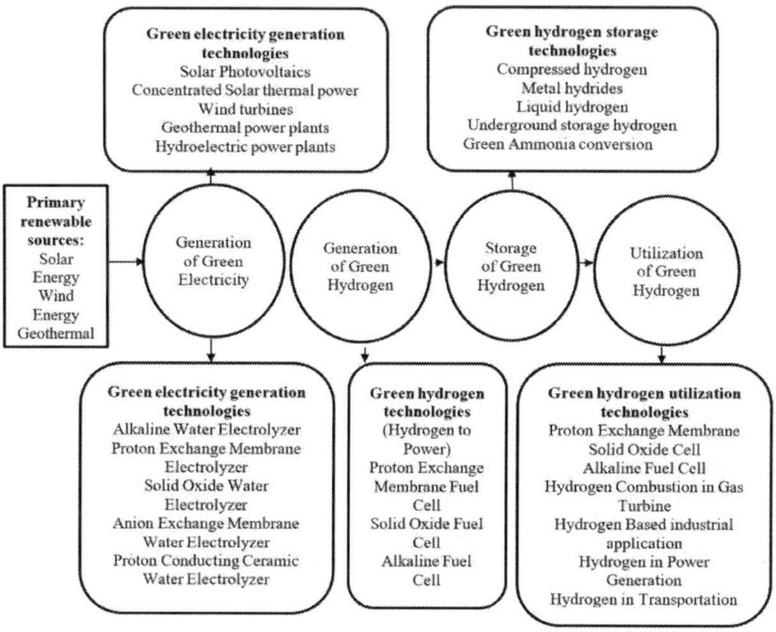

Figure 9.1. *Schematic representation of the core components of a green hydrogen energy system*

The fundamental operating principle of a GHES begins with harnessing renewable energy sources such as solar, wind, hydropower and geothermal energy. The generation of electricity using technologies such as WTs, solar PV modules, concentrated solar thermal power plants, hydroelectric generators and geothermal power plants is commonly called "green electricity" – a term that denotes production without the use of fossil fuels or using an environmentally friendly process. A critical indicator for assessing the technoeconomic viability of the GHES is the levelized cost of electricity. This assessment represents the average net present cost of electricity generation for a precise technology over its entire operational lifespan. The levelized cost is used as a key tool for analyzing and comparing different green electricity generation methods as specified in Table 9.2.

RES	Installed capacity (GW)	Share in electricity production (%)	Total Installed costs (₹/kW)	Annual O&M costs (₹/kW)	Capacity factor (%)	LCOE (₹/kWh)
Solar PV	81.8	7.2	67,000–74,000	300– 12,000	19	3.5
Onshore wind	45.7	3.7	95,000–110,000	800– 1,500	23	2.8
Offshore wind	0 (planned 30 by 2030)	0	175,000 (est.)	2,500 (est.)	35 (est.)	5.5 (est.)
CSP	<1	<0.1	3,000,000 (est.)	4,000 (est.)	20 (est.)	9.0 (est.)
Hydropower	42.1	11.2	80,000–120,000	1,500–4,000	45	4
Geothermal	0 (pilot Projects only)	–	–	–	–	–

Table 9.2. *Indian green electricity production overview (2023-2024): installed capacity, production share, global weight average values for installed costs, capacity factors and levelized cost of electricity (LCOE) in various electricity generation technologies (IRENA 2020; IEA 2021)*

NOTE.– Currency converted from global estimates using the average exchange rate: 1 USD ≈ ₹83 (2023–2024 average). LCOE values are approximate and vary regionally within India. Offshore wind, CSP and geothermal are not yet significantly deployed in India but are included for projection purposes.

9.2.2. *Production of green hydrogen*

The core principle behind hydrogen production involves two essential components: a feedstock (from which hydrogen is derived) and an energy source to derive the extraction process. In green hydrogen systems, the production of green electricity sources powers water electrolyzers that split water molecules to release hydrogen. This process is a subset of the broader Power-to-X concept, where "X" refers to the desired energy carrier or product as in our case of green hydrogen (International Renewable Energy Agency (IRENA) 2022). In a standard water electrolysis cell, water is split using an electrolyte (which conducts ions) and two electrodes (catalyst-coated electronic conductors): the cathode and the anode. During this process, hydrogen gas is produced at the cathode through the hydrogen evolution reaction, whereas the oxygen is generated at the anode via the oxygen evolution reaction (Tahmasbi et al. 2025). Various electrolysis technologies exist, each defined by different types of electrolytes, separators and operating conditions. Table 9.3 lists the materials commonly used in different water electrolysis technologies. Table 9.4 presents the strengths, drawbacks and current challenges of each method.

Component	AWE	PEMWE	SOWE	AEMWE	PCWE
Electrolyte-separator	Aqueous KOH (20–40 wt%) or NaOH includes stabilizers like PTFE	Perfluorosulfonic acid (PFSA) membranes and advanced ion-conductive polymers	YSZ, GDC	AEP, DVB polymer with KOH, $NaHCO_3$	ABO-based perovskites with KOH, LaBO-phosphate
Electrocatalyst anode	Nickel-based stainless-steel mesh	IrO_2, RuO_2, mixed oxides	Perovskite structured materials (LSC)	High surface Ni, Fe, Co alloys	Mixed oxide perovskites (MIEC, MPEC)
PTL anode	Ni mesh (optional)	Titanium mesh or foam with platinum coating	Coarse porous structures like nickel foam	Ni foam	Nickel mesh with adequate porosity
Bipolar Plate anode	Stainless steel with nickel plating	Titanium with gold or conductive coatings	Nickel-based or oxidized composites	Steel plates with nickel coating	Steel with corrosion-resistant surface treatment
Electrocatalyst cathode	Nickel-rich materials	Platinum group metal-based electrocatalysts	Mixtures like Ni-YSZ or Ni-GDC	Nickel or cobalt-based catalysts	Nickel-ceramic composites such as Ni-BZY or Ni-LSC
PTL cathode	Nickel mesh	Carbon or titanium felt	Nickel-based porous supports	Similar to anode–carbon/Ni foams	Conductive porous foams
Bipolar plate cathode	Same as anode – Ni-plated steel	Titanium with coating	Nickel or stainless-steel variants	Ni-coated stainless steel	Treated steel or metal-ceramic blends
Frames and sealing	Polymeric seals (e.g. PTFE and EPDM)	PTFE, PSU, ETFE	Glass, silica-based materials	PTFE, silicon	Ceramic glass

Table 9.3. *Materials commonly used in different water electrolysis technologies*

	Strengths	Drawbacks/challenges
AWE	– Produces hydrogen with extremely high purity (typically above 99.999%) – Functions efficiently at room temperature conditions – Offers extended operational lifespan with high durability – Utilizes cost-effective catalysts without the need for precious metals – Lower investment and manufacturing costs due to simple system design	– Generally limited to lower current densities – Corrosion risks due to the presence of liquid alkaline electrolyte – Formation of gas bubbles can disrupt ion movement and raise electrical resistance – Risk of hydrogen and oxygen mixing affects system safety and hydrogen purity – Power output per unit area is relatively lower
PEMWE	– High current output (up to 10 A/cm²) – Compact design – Produces high-purity hydrogen – Dynamic performance and fast response – Integrates with renewables – Operates up to 700 bar without external compression	– High cost of catalyst (PGM) – Acidic conditions cause electrode degradation – Needs water purification – Pressurization and component lifespan are key concerns
SOWE	– Enhanced reaction rates and thermodynamics – Effective heat management – Suitable for reversible operation (fuel/electrolyzer) – High scalability – Lower potential capital cost	– Corrosion issues reduce lifespan – High start-up/shut-down times – Products contain steam; cooling and compression needed – Susceptible to performance loss
AEMWE	– Combines benefits of AWE and PEMWE – Improved efficiency and kinetics – Operates in an alkaline medium and is less corrosive – Reduced nickel oxidation and leaching	– Anion exchange membranes face stability issues – Mechanical strength is lower than PEMWE – Faradaic losses – Currently high cost and scaling challenges

Table 9.4. *Strengths, drawbacks and present challenges of each method*

9.2.3. *Storage and utilization of green hydrogen*

The storage and use of hydrogen are important parts of an end-to-end GHES that render it reliable, flexible and economical. As green electricity is produced based on renewable energy sources such as solar, wind, hydro or geothermal energy, storing hydrogen acts as the vital intermediate between the supply of irregularly fluctuating energy demand and the generation of intermittent energy. This storage role assists in stabilizing the energy system and offers the capacity to transfer energy consumption over time and sectors, which is crucial for decreasing the dependence on fossil fuels and reducing end-user costs. Hydrogen can be stored either physically or chemically, the best established means of which are compressed gaseous hydrogen, which is simple and well established, underground storage in geological structures such as salt caverns or aquifers, which can support large-scale storage, and metal hydride storage, which involves absorbing hydrogen in solid metal alloys with high volumetric density. In addition, other potential options involve liquid hydrogen (LH_2) storage with high energy density but at the expense of being stored at cryogenic temperatures, and green ammonia synthesis in which hydrogen is stored and transported chemically as ammonia to be reconverted on demand.

Storage method	State	Typical pressure/temp.	Volumetric energy density (MJ/L)	Gravimetric energy density (wt%)	Efficiency (%)
Compressed hydrogen gas	Gas	350–700 bar, ambient temp.	~5–8 MJ/L	~4–5 wt%	40–55
Liquid hydrogen (LH_2)	Liquid	~20 K, ~1 atm	~8.5 MJ/L	~7–8 wt%	25–45
Metal hydrides	Solid	10–100 bar, 300–600 K	1–5 MJ/L	1–7 wt%	50–75
Chemical hydrogen storage	Solid/liquid	Mild to moderate conditions	2–6 MJ/L	~10 wt%	30–50
Liquid organic hydrogen carriers (LOHCs)	Liquid	Ambient to mild conditions	5–7 MJ/L	5–6 wt%	30–50
Underground storage (e.g. salt caverns)	Gas or compressed	100–300 bar, ambient	Site dependent	Site dependent	High

Table. 9.5. *Comparison of hydrogen storage methods and performance parameters*

Every storage mode has its specific technical and financial attributes, for instance, pressure and temperature to operate, storage efficiency, storage density and levelized cost of hydrogen storage (LCHS). The choice of a suitable storage technology is, to a great extent, based on the targeted application case – ranging from mobile and fixed energy applications to industrial and transportation industries – and on other factors such as infrastructure availability, safety, cost-effectiveness and environmental considerations, including emissions related to manufacturing and deployment of storage devices. As the technologies develop, their application within energy systems will be increasingly determined by the kind of electrolyzer employed to produce hydrogen, the availability of resources and local technological advancement. A detailed comparison of the reported hydrogen storage methods is available in Table 9.5.

After storage, the use of green hydrogen is usually electricity, allowing it to power houses, vehicles, industries and whole communities. This is done mainly by fuel cells, which are very efficient electrochemical converters that generate electricity by the controlled reaction of hydrogen and oxygen, with water and heat as the sole by-products. In contrast to combustion engines, fuel cells are quiet, have no moving parts and provide much higher energy conversion efficiency. There are different types of fuel cells that have been classified based on their electrolytes, mode of ionic transport and operating conditions. These are proton exchange membrane fuel cells (PEMFCs) – ideal for transportation and portable applications due to their fast start-up and low operating temperature; solid oxide fuel cells (SOFCs) – ideal for stationary power generation due to their high efficiency and fuel flexibility; alkaline fuel cells (AFCs); phosphoric acid fuel cells (PAFCs); and molten carbonate fuel cells (MCFCs). Each of them has specific benefits and trade-offs in cost, durability, efficiency and scalability. The general electrochemical reactions that govern different forms of fuel cells are listed in Table 9.6. There has been a recent improvement, which has led to new-emerging technologies including anion exchange membrane fuel cells (AEMFCs), which propose to integrate the advantages of both AFC and PEM fuel cells, and proton-conducting ceramic fuel cells (PCFCs), operating at intermediate temperature and with guaranteed higher efficiency and longer lifespan. These new systems are quickly gaining attention because of their potential to lower costs and enhance performance. Ultimately, the effective integration of green hydrogen storage and utilization technologies creates a closed-loop cycle – power-to-green hydrogen-to-power – that has the potential to transform the world's energy landscape by offering a sustainable, clean and flexible alternative to traditional fossil fuels.

	Anode	**Cathode**
AFC (AEMFC)	$H2 + 2OH^- \rightarrow 2H_2O + 2e^-$	$1/2\ O_2 + 2e^- + H_2O \rightarrow 2OH^-$
PEMFC	$H_2 \rightarrow 2H^+ + 2e^-$	$1/2\ O_2 + 2H^+ + 2e^- \rightarrow H_2O$
PAFC	$H_2 \rightarrow 2H^+ + 2e^-$	$1/2\ O_2 + 2H^+ + 2e^- \rightarrow H_2O$
MCFC	$H_2 + CO_3^{2-} \rightarrow CO_2 + H_2O + 2e^-$	$1/2\ O_2 + CO_2 + 2e^- \rightarrow CO_3^{2-}$
SOFC	$H_2 + O^{2-} \rightarrow H_2O + 2e^-$	$1/2\ O_2 + 2e^- \rightarrow O^{2-}$
PCFC	$H_2 \rightarrow 2H^+ + 2e^-$	$1/2\ O_2 + 2H^+ + 2e^- \rightarrow H_2O$

Table. 9.6. *Electrochemical conversion mechanisms*

9.3. Integrated modeling, experimental investigations and applied projects aimed at advancing green hydrogen energy solutions

Integrated modeling plays a critical role in evaluating the scalability, economic viability and practicality of green hydrogen technologies. These models combine technical parameters – such as electrolyzer efficiency, power consumption and hydrogen storage characteristics – with economic factors including capital expenditure (CAPEX), operating expenditure (OPEX), electricity costs and market dynamics. *Techno-economic models* are often developed to focus on optimal *system configurations* for green hydrogen production, with the primary objective of *minimizing the LCOH*. More advanced modeling approaches also incorporate lifecycle assessment (LCA) and sensitivity analyses to address uncertainties in market evolution, policy impacts and technological advancements. These tools offer valuable insights for decision-making in policy, investment and infrastructure planning. Several studies have compared electrolyzer types based on their environmental and operational performance.

For instance, Lundberg and Sundin found that proton exchange membrane water electrolysis (PEMWE) had the lowest environmental impact compared to alkaline water electrolysis and solid oxide water electrolysis using 100 kg of hydrogen as a benchmark. This was primarily due to PEMWE's low energy consumption and lightweight design. However, they also emphasized that the environmental footprint is heavily influenced by the electrolyzer's operational lifespan and the current density. Zhao et al. conducted a lifecycle inventory (LCI) analysis to quantify material and energy requirements for producing 1 m^2 of electrolysis cells. They applied the lifecycle impact assessment (LCIA) methodology, specifically the ReCiPe 2016 framework, to evaluate impacts across 16 categories, including climate change and resource depletion. Their results showed that PEMWE stacks have a higher environmental footprint than SOWE and AWE stacks, particularly in global warming potential and mineral resource depletion. Gerloff's research indicated that SOWE is the least environmentally damaging under current and near-term energy systems, while PEMWE is better suited for future systems based entirely on renewable energy by 2050. Khan et al. underscored the importance of conducting full LCAs of the entire electrolyzer supply chain in accordance with ISO standards. Their findings demonstrated that localized production and recycling practices could reduce the environmental impact of green hydrogen by up to 20%. Vilbergsson and colleagues performed a cradle-to-gate LCA of hydrogen production in Austria, Belgium and Iceland. They assessed how different electrolyzer types, energy sources and operating conditions affect the overall carbon footprint. Interestingly, their analysis revealed that transporting hydrogen from Iceland to mainland Europe had a relatively minor environmental impact compared to other operational factors. Additionally, there has been a growing body of techno-economic and experimental research on GHES, as reflected in the expanding number of projects listed in the IEA 2023 hydrogen project database.

Figure 9.1 illustrates that nearly 90% of hydrogen production projects are still in the planning phase, with only a small fraction currently operational. Figures 9.2(a) and (b) detail the stages of renewable hydrogen projects – ranging from planning and construction to deployment – whereas Figure 9.3 categorizes supporting infrastructure projects, such as pipelines and storage facilities, by their development status. Although many projects do not explicitly state their energy sources, solar and wind are among the most widely utilized. Concretized technical and economic factors impacting the cost of hydrogen are clearly classified in Table 9.7.

Performance metrics	**Cost metrics**		
	Capital investment	**Operating cost**	**Another payback period**
– System size – Efficiency and electricity use – Efficiency at partial load – Availability (capacity factor) – Operating load range compared to rated capacity – Efficiency drops overtime (degradation) – Cell durability – Thermal and water demand – H_2 concentration level – Startup times (warm and cold)	– Cost of electrolyzer system – Cost of supporting equipment (BoP) – Construction (land, buildings and foundation) – Electrical network linkage – Fuel storage module – Mobility infrastructure – Savings from scaling up – Plant development cost	– Maintenance of system and components – Replacement of stack and BoP – Electricity and water feedstock costs (fresh, recycled and desalinated) – Savings from scaling up (scaling factor) – Earnings from by-products (oxygen and heat) – Logistics, labor, insurance and overhead costs	– Time value of money rates – Taxation and grant schemes – Power supply rates – Capital structure components – Evaluation timeframe – Terminal value – Budget buffer – Government or business incentives – Decommissioning costs

Table. 9.7. *Breakdown of technical and economic parameters impacting hydrogen cost*

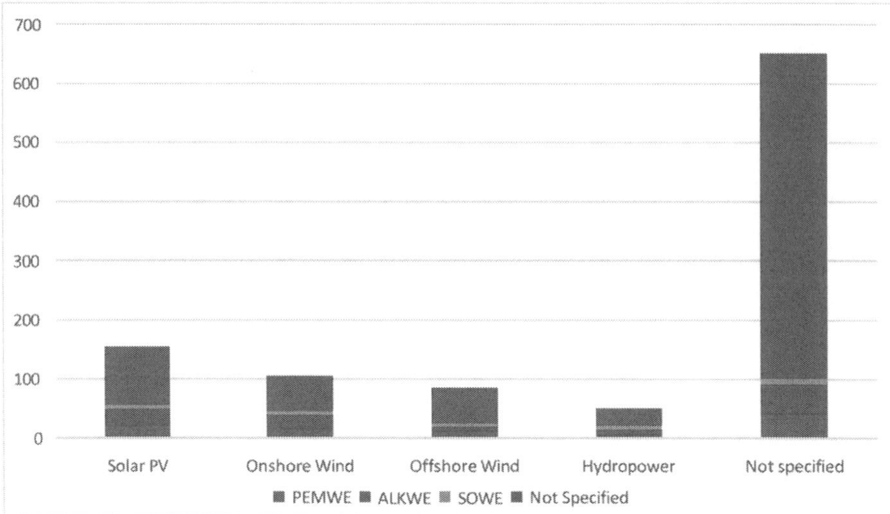

a)

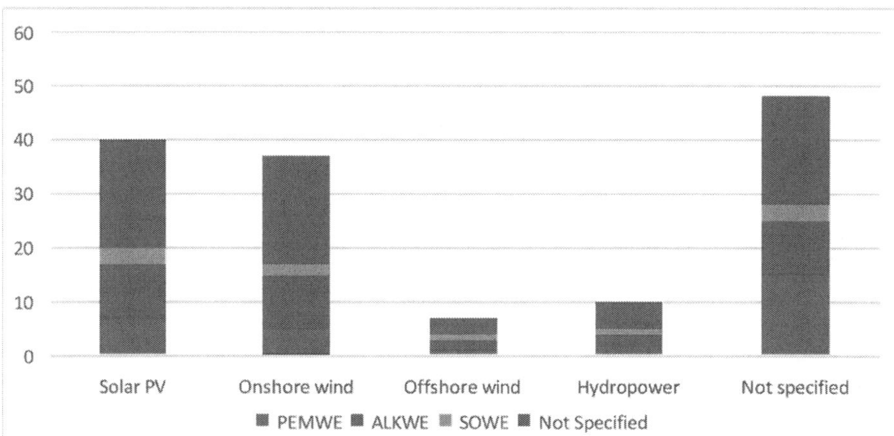

b)

Figure. 9.2. *Breakdown of electrolysis-based hydrogen projects using only RES: (a) total projects by development stage and (b) projects in operation. The term "not specified" applies when the exact renewable energy input or its configuration is not detailed. For a color version of this figure, see www.iste.co.uk/sundaram/energysystems.zip*

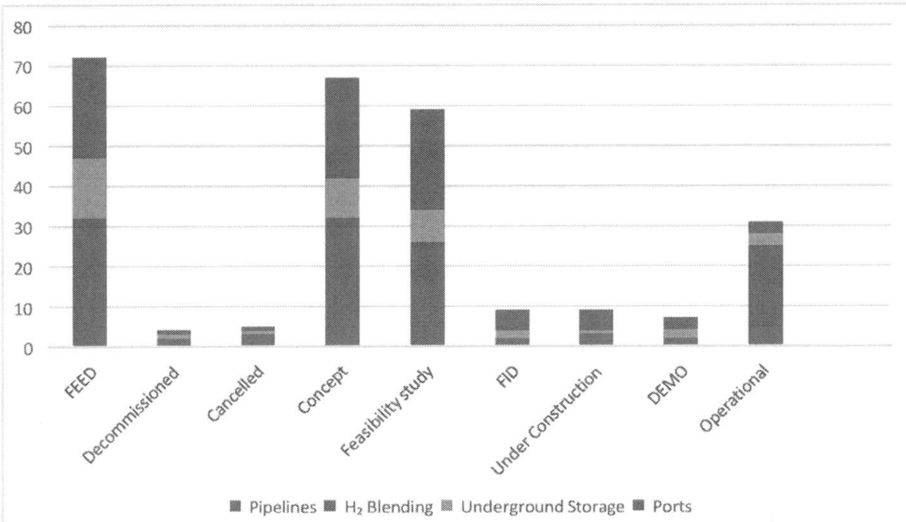

Figure 9.3. *Categorization of hydrogen infrastructure projects by type and lifecycle stage. For a color version of this figure, see www.iste.co.uk/sundaram/energysystems.zip*

9.3.1. *Sunlight-driven energy systems*

9.3.1.1. *Direct PV energy conversion*

The concept of using sunlight to produce hydrogen has been around since the mid-20th century, when researchers began exploring electrochemical processes for water splitting using renewable energy. Early systems were largely experimental, hindered by low-efficiency solar cells, high costs and the lack of established electrolyzer technology. The oil shocks of the 1970s brought alternative energy sources into the spotlight, accelerating interest in solar hydrogen systems. However, progress remained slow due to both economic constraints and technical limitations. These initial systems typically relied on simple PV panels connected to basic electrolyzers, with low efficiency and minimal integration. The high cost of solar panels during that era made such systems economically viable only for laboratory research or niche applications. Nonetheless, this period laid important groundwork for future advancements in solar cell and water-splitting technologies. In recent years, there have been transformative advances in both PV and electrolyzer technologies, significantly enhancing the viability of PV-powered hydrogen production. The cost of solar PV systems has dropped dramatically, while the efficiency and durability of both solar panels and electrolyzers have improved. Today's systems typically use PEM or alkaline electrolyzers that can operate effectively with variable power inputs from solar arrays. Green hydrogen production through solar PV is now realized via two main approaches: direct integration, where PV panels are directly connected to electrolyzers without significant power conditioning, and indirect integration, where solar power feeds into the grid or is stored before being used to power electrolyzers. Future systems are expected to incorporate AI-based optimization, smart grid connectivity and large-scale deployment. With continued policy support, technological innovation and cost reductions, PV-based hydrogen production is poised to become a key pillar in the global transition to a decarbonized energy system.

9.3.1.2. *Thermal solar power applications*

Thermal solar power applications focus sunlight onto a small area to create electricity using heat, making it a good option for generating green hydrogen, especially in places that get a lot of sun all year round. A few studies have looked into how well thermal solar power can make hydrogen, checking things like its efficiency and how much it costs to produce it. While there has been a lot of research on this topic, we have not seen many real-world examples of thermal solar power hydrogen projects. One notable case is a recent project in California by Bloom Energy and Hydrogen, where they used a solar tower with solid oxide water electrolysis to create green hydrogen.

9.3.1.3. *Water energy-based applications*

According to data from the IEA (2021) hydrogen database, there are currently 43 GHES projects powered by hydropower, ranging from early-stage concepts to fully operational facilities – of which seven are currently active, as shown in Figure 9.1. One notable example is a collaborative project launched in Switzerland in 2020, where an electrolysis plant was installed at the Alpiq Gösgen hydropower site. This facility uses a 2 MW PEMWE and is capable of producing around 300 tons of green hydrogen annually (Hydrogen Council 2021; Ling et al. 2021). The hydrogen is intended primarily for fuelling heavy-duty fuel cell trucks and passenger fuel cell vehicles (Staffell et al. 2019; Roy et al. 2020). Another significant project is the "REMOTE" initiative, funded by the EU's Horizon 2020 program. Its goal was to develop advanced power-to-green hydrogen-to-power systems tailored for remote and off-grid regions. One of its demonstration sites, known as DEMO 2, went live in 2021 in Agkistro, Greece. The setup includes a 0.9 MW hydropower unit, a 25 kW alkaline water electrolyzer (AWE) and a 50 kW PEMFC. The system also integrates a hydrogen storage unit (HSU) capable of holding 30 kg at 30 bar and a 30 kWh lithium-ion battery, offering a complete and autonomous clean energy solution

9.3.1.4. *Hydrogen from geothermal sources*

Green hydrogen production powered by geothermal energy is gaining traction worldwide due to its low operating costs and impressive efficiency (Dincer and Acar 2015; Brahim and Jemni 2026). While many technical studies highlight the potential of combining these two sustainable technologies, real-world implementations have been relatively limited. However, Iceland stands out as a leader in this space. Reykjavik Energy has launched the country's first commercial geothermal green hydrogen facility, using 750 kW of geothermal power to run advanced AWEs from NEL. In New Zealand, the green hydrogen movement picked up pace in 2021 with the opening of a 1.5 MW facility in Taupo by Halcyon Power. This facility, fueled by electricity from the adjacent Mokai geothermal power station, employs PEMWEs to generate up to 250 Nm^3 of hydrogen per hour, supplying both the transportation and energy sectors (Roy et al. 2020; Hydrogen Council 2021; Shen et al. 2024). Building on this achievement, the same team recently launched a pilot project in Japan's Oita Prefecture. This smaller scale project combines a 125 kW binary geothermal power system with hydrogen production equipment, demonstrating how even limited geothermal resources can support clean energy initiatives.

9.3.1.5. *Integration of multiple energy sources*

These strategies integrate various forms of renewable electricity generation – such as PV, concentrated solar power (CSP), WTs, hydropower and geothermal energy. Although combining multiple sources can result in complex systems, careful design and management can effectively leverage the synergies among these technologies (Gallo et al. 2016; Wang et al. 2017).

By combining the strengths of different energy systems, multi-energy-based solutions offer a powerful way to improve overall efficiency and reliability. The variability in renewable energy availability, however, can significantly impact the system's performance. For example, when electrolyzers and fuel cells in GHESs operate at partial loads, it can lead to faster wear and tear compared to consistent, full-load operation (Buttler and Spliethoff 2018; Ling et al. 2021). Additionally, if the system is not correctly sized to handle these fluctuations, it can affect reliability. This challenge also drives up capital costs, as larger electrolyzers, fuel cells and HSUs may be necessary to manage these variations.

PV systems are some of the most extensively researched technologies for hydrogen production. Concentrated photovoltaic (CPV) systems paired with electrolysis have demonstrated solar-to-hydrogen (STH) conversion efficiencies reaching up to 28%. Large-scale PV-electrolyzer systems could potentially lower the cost of hydrogen to as little as ₹122.55 and ₹143.97 per kilogram. The simplest way to link PV systems with electrolyzers is through direct coupling, which helps stabilize hydrogen production and extend the system's operational hours. These systems can benefit from grid support or energy storage-assisted electrolysis, such as batteries. However, while batteries improve the utilization of PV systems, they also increase the overall energy cost due to their upfront investment, maintenance and replacement needs, as discussed by Martinez et al. (2023).

Direct PV energy conversion offers another promising method for hydrogen production, with efficiencies of around 26% and the potential for a low LCOH of ₹235.675 per kilogram. By integrating thermal energy storage (TES), the system can run longer and more reliably, ensuring a consistent supply of hydrogen.

Thermal solar power-based systems also show great promise for cost-effective hydrogen production, with projected LCOH values ranging from ₹59.99 to ₹342.8 per kilogram, depending on location. Similar to PV systems, intermittency challenges can be managed using short-term storage solutions like batteries.

Water energy-based systems can leverage existing infrastructure to produce hydrogen at a low cost, with LCOH values ranging from ₹102.84 to ₹342.8 per kilogram.

Geothermal energy is another strong contender for hydrogen production, offering high efficiencies and low LCOH of approximately ₹94.27–₹342.8 per kilogram.

9.4. A thorough review and exploration of future opportunities for GHESs

An exhaustive and critical assessment of GHESs identifies their potential to transform the global energy market, particularly with regard to reaching profound decarbonization goals. Green hydrogen, as hydrogen generated from the electrolysis of water employing renewable electricity, is among the most promising energy carriers of the future because it can fill the gap between discontinuous renewable generation and the continuous demand for energy. In contrast to blue or grey hydrogen, which are fossil-fuel-based and carry carbon emissions, green hydrogen offers a clean and sustainable option, especially for industries where direct electrification is economically or technically not possible – like steel production, chemical industries, aviation, shipping and industrial-scale power generation. The critical analysis emphasizes incorporating green hydrogen into the larger energy system, highlighting the ability of green hydrogen to increase the stability of the grid, provide seasonal energy storage and serve as fuel for fuel-cell-based transportation and industrial applications.

Technologically, the GHES environment is changing fast with the development of several electrolysis technologies and ongoing advancements.

AWEs, PEMWE, SOWE and next-generation technologies like anion exchange membrane electrolyzers (AEMWE) and proton-conducting ceramic electrolyzers (PCWE) provide different levels of efficiency, design flexibility, material sustainability and integration capability with renewable energy. Each technology possesses unique strengths and weaknesses; for example, AWE systems are mature and low-cost but do not possess the dynamic responsiveness of PEMWE systems, which are better suited to variable renewable input but rely on costly noble metals. Likewise, high-temperature electrolyzers such as SOWEs provide improved efficiency through heat integration but are plagued by material degradation and complicated system requirements. Future development projects must thus seek to strike a balance between cost, performance, longevity and environmental footprint among these technologies. Hydrogen storage and supply are also key aspects of GHES. Efficient storage technologies such as compressed and liquefied hydrogen, underground geological storage and chemical carriers such as ammonia are crucial to ensuring hydrogen becomes a stable part of current energy networks. However, each has trade-offs in energy density, infrastructure cost, safety and environmental footprint. Establishing strong, scalable and cost-effective storage infrastructure will be central to unlocking green hydrogen's full potential, particularly when combined with fuel cells for electricity or transportation. Additionally, the combination of green hydrogen with current energy grids and industrial systems necessitates smart energy management systems, AI-driven optimization and digital control structures to enable stable, secure and efficient operations under dynamic load and supply conditions. Economically and from a policy standpoint, the green hydrogen value chain is presently limited by expensive production, missing infrastructure and dispersed regulatory systems. The LCOH is still high in relation to alternatives from fossil fuels, although it is expected to fall with technological advancements, economies of scale and supporting market mechanisms. Governments, industries and research organizations need to act together to develop innovation, standardization and risk-sharing arrangements through public–private partnerships, incentives and subsidies. International cooperation will also be necessary to harmonize safety requirements, trade practices and certification schemes for hydrogen purity and sustainability. In addition, LCAs and technoeconomic modeling will have to be used to inform decision-making and invest in those priorities that yield maximum environmental and economic returns. In the coming years, green hydrogen systems are expected to become multi-energy hybrid platforms integrating solar, wind, hydro and geothermal energy sources with smart grids and storage technologies. The hybrid systems will be able to offer decentralized and reliable energy solutions, particularly in off-grid or rural areas. Predictive maintenance enabled by artificial intelligence, real-time optimization of the system and the principles of modularity will make the systems even more efficient and lower their cost of operation. As materials science, electrochemistry and digital engineering innovations continue to evolve, green hydrogen may not only be an energy complement but also a keystone of the global clean energy economy. In short, though today's challenges are substantial, the future of GHES is promising, as long as there is a continued dedication across scientific, political and industrial sectors to develop innovation, decrease costs and create the infrastructure needed for a hydrogen-dominated world.

9.5. Conclusion

GHESs are at the center of the world's shift to a clean, robust and sustainable energy economy. As a zero-carbon fuel that is generated from renewable power, green hydrogen presents an influential answer to decarbonize those sectors that are otherwise hard to electrify, such as heavy industry, long-distance transportation and large power generation. In this review, each of the component technologies and system components that make up GHES has been reviewed critically, including renewable electricity supply and high-end electrolyzer architectures as well as hydrogen storage strategies and fuel cell use – discussing not only their capability now but their current limitations too. Although technical advancement and installation are well along the way toward solutions, considerable problems remain about high production prices, sparse infrastructure, system complexity for integration and the imposition of standardized standards. Overcoming these challenges demands a collaborative and multi-disciplinary approach through research and innovation, supportive policies and strategic investments. Quantifiable targets such as improving electrolyzer efficiency beyond 75%, reducing dependence on rare materials by 50%, cutting storage energy losses below 10% and achieving green hydrogen production costs under $2/kg by 2030 are essential to drive commercialization and market scale-up. Development in the future needs to emphasize increasing electrolyzer efficiency, lowering reliance on rare materials, advancing storage solutions and using AI and smart grid technologies for maximizing system performance. Eventually, green hydrogen has the potential to emerge as a linchpin of the global energy mix, allowing for a low-carbon economy and assisting with climate ambitions, energy security and industrial competitiveness. Through ongoing dedication by governments, industries and academia, GHESs can transform from developing technology to common place, powering a cleaner, greener world.

9.6. References

2035 Energy Mix Study Team, Renewable Energy Institute (2024). Energy transition scenarios for deep decarbonization. Energy Strategy Reviews [Online]. Available at: https://www.renewable-ei.org/pdfdownload/activities/REI_2035study_EN_2408.pdf.

Ajeeb, W., Neto, R.C., Baptista, P. (2024). Life cycle assessment of green hydrogen production through electrolysis: A literature review. *Sustainable Energy Technologies and Assessments*. 69, 103923. doi: 10.1016/j.seta.2024.103923.

Akinyele, D., Olabode, E., Amole, A. (2020). Review of fuel cell technologies and applications for sustainable microgrid systems. *Inventions*, 5(3), 42. doi: 10.3390/inventions5030042.

Ball, M. and Weeda, M. (2015). The hydrogen economy – Vision or reality? *International Journal of Hydrogen Energy*, 40(25), 7903–7919.

Bhandari, R. et al. (2014). Life cycle assessment of hydrogen production methods: A review. *Journal of Cleaner Production*, 85, 151–163.

Bicer, Y. and Dincer, I. (2018). Clean fuel options with hydrogen for sea transportation: A techno-economic and environmental assessment. *International Journal of Hydrogen Energy*, 43(2), 1179–1193.

Boretti, A. and Pollet, B.G. (2024). Hydrogen economy: Paving the path to a sustainable, low-carbon future. *International Journal of Hydrogen Energy*, 93, 307–319. doi: 10.1016/j.ijhydene.2024.10.350.

Brahim, T and Jemni, A. (2026). Green hydrogen production: A review of technologies, challenges, and hybrid system optimization. *Renewable and Sustainable Energy Reviews*, 225, 116194. doi: 10.1016/j.rser.2025.116194.

Buttler, A. and Spliethoff, H. (2018). Current status of water electrolysis for energy storage, grid balancing and sector coupling via power-to-gas and power-to-liquids. *Renewable and Sustainable Energy Reviews*, 82, 2440–2454.

Chen, Z., Song, P., Wang, B. (2021). Carbon emissions trading scheme, energy efficiency and rebound effect – Evidence from China's provincial data. *Energy Policy*, 157, 112507. doi: 10.1016/j.enpol.2021.112507.

Colmenar, J.M., Hidalgo, J.I., Salcedo-Sanz, S. (2018). Automatic generation of models for energy demand estimation using grammatical evolution. *Energy*, 164, 183–193. doi: 10.1016/j.energy.2018.08.199.

Dincer, I. and Acar, C. (2015). Review and evaluation of hydrogen production methods for better sustainability. *International Journal of Hydrogen Energy*, 40(34), 11094–11111.

European Commission (2020). A Hydrogen Strategy for a Climate-Neutral Europe. European Commission, Brussels.

Gahleitner, G. (2013). Hydrogen from renewable electricity: An international review of power-to-gas pilot plants for stationary applications. *International Journal of Hydrogen Energy*, 38(5), 2039–2061.

Gallo, A.B. et al. (2016). Energy storage in the energy transition context: A techno-economic review. *Renewable and Sustainable Energy Reviews*, 65, 799–823.

Gerloff, N. (2021). Comparative Life-Cycle Assessment analysis of power-to-methane plants including different water electrolysis technologies and CO_2 sources while applying various energy scenarios. *ACS Sustainable Chemistry & Engineering*, 9(30) 10123–10141. doi: 10.1021/acssuschemeng.1c02002.

Glenk, G. and Reichelstein, S. (2019). Economics of converting renewable power to hydrogen. *Nature Energy*, 4(3), 216–222.

Greene, D.L., Ogden, J.M., Lin, Z. (2020). Challenges in the designing, planning and deployment of hydrogen refueling infrastructure for fuel cell electric vehicles. *eTransportation*, 6, 100086. doi: 10.1016/j.etran.2020.100086.

Huang, B.S., Chen, H.Y., Chuang, K.H., Yang, R.X., Wey, M.Y. (2012). Hydrogen production by biomass gasification in a fluidized-bed reactor promoted by an Fe/CaO catalyst. *International Journal of Hydrogen Energy*, 37(8), 6511–6518. doi: 10.1016/j.ijhydene.2012.01.071.

Hydrogen Council (2021). Hydrogen Insights – A Perspective on Hydrogen Investment, Deployment and Cost Competitiveness. Report, Hydrogen Council, Brussels.

International Energy Agency (IEA) (2019). The Future of Hydrogen: Seizing Today's Opportunities. Report, OECD, Paris. doi: 10.1787/1e0514c4-en.

International Energy Agency (IEA) (2021). Hydrogen Projects Database. IEA, Paris.

International Energy Agency (IEA) (2022). Global Hydrogen Review 2022. Press Release, IEA, Paris.

International Renewable Energy Agency (IRENA) (2020). *Green Hydrogen Cost Reduction: Scaling up Electrolysers to Meet the 1.5°C Climate Goal*. IRENA, Abu Dhabi.

International Renewable Energy Agency (IRENA) (2022). *Green Hydrogen: A Guide to Policy Making*. IRENA, Abu Dhabi.

Janssen, J.L.L.C.C., Weeda, M., Detz, R.J., van der Zwaan, B. (2022). Country-specific cost projections for renewable hydrogen production through off-grid electricity systems. *Applied Energy*, 309, 118398. doi: 10.1016/j.apenergy.2021.118398.

Khan, M.A. (2021). Demonstration of green hydrogen production using solar energy at 28% efficiency and evaluation of its economic viability. *Sustainable Energy Fuels*, 2021(5), 1085–1094. doi: 10.1039/D0SE01761B.

Kong, X., Ma, L., Wang, C., Guo, S., Abdelbaky, M.A., Liu, X., Lee, K.Y. (2022). Large-scale wind farm control using distributed economic model predictive scheme. *Renewable Energy*, 181, 581–591. doi: 10.1016/j.renene.2021.09.048.

Li, Q., Molina Villarino, A., Peltier, C.R., Macbeth, A.J., Yang, Y., Kim, M.J., Shi, Z., Krumov, M.R., Lei, C., Rodríguez-Calero, G.G. et al. (2023). Anion exchange membrane water electrolysis: The future of green hydrogen. *The Journal of Physical Chemistry C*, 127(17), 7901–7912. doi: 10.1021/acs.jpcc.3c00319.

Ling, Y., Kazim, F.M.D., Zhang, Q., Xiao, S., Li, M., Yang, Z. (2021). Construction of Mo2C/W2C heterogeneous electrocatalyst for efficient hydrogen evolution reaction. *International Journal of Hydrogen Energy*, 46(15), 9699–9706. doi: 10.1016/j.ijhydene.2020.06.291.

Lototskyy, M.V., Yartys, V.A., Pollet, B.G., Bowman Jr., R.C.. (2014). Metal hydride hydrogen compressors: A review. *International Journal of Hydrogen Energy*, 39(11), 5818–5851.

Martinez Lopez, V.A., Ziar, H., Haverkort, J.W., Zeman, M., Isabella, O. (2023). Dynamic operation of water electrolyzers: A review for applications in photovoltaic systems integration. *Renewable and Sustainable Energy Reviews*, 182, 113407. doi: 10.1016/j.rser.2023.113407.

Ministry of New and Renewable Energy (MNRE), Government of India (2023). National Green Hydrogen Mission. MNRE, Government of India, New Delhi.

Muthukumar, P., Kumar, A., Afzal, M., Bhogilla, S., Sharma, P., Parida, A., Jana, S., Kumar, E.A., Pai, R.K., Jain, I.P. (2023). Review on large-scale hydrogen storage systems for better sustainability. *International Journal of Hydrogen Energy*, 48(85), 33223–33259. doi: 10.1016/j.ijhydene.2023.04.304.

Nielsen, K.S., Nicholas, K., Creutzig, F., Dietz, T., Stern, P. (2021). The role of high-socioeconomic-status people in locking in or rapidly reducing energy-driven greenhouse gas emissions. *Nature Energy*, 6. doi: 10.1038/s41560-021-00900-y.

Nikolaidis, P. and Poullikkas, A. (2017). A comparative overview of hydrogen production processes. *Renewable and Sustainable Energy Reviews*, 67, 597–611.

Okonkwo, P.C., Islam, M.S., Taura, U.H., Barhoumi, E.M., Mansir, I.B., Das, B.K., Ali Sulaiman, M.M.B., Agyekum, E.B., Bahadur, I. (2024). A techno-economic analysis of renewable hybrid energy systems for hydrogen production at refueling stations. *International Journal of Hydrogen Energy*, 78, 68–82. doi: 10.1016/j.ijhydene.2024.06.294.

Olugasa, T.T., Odesola, I.F., Oyewola, M.O. (2014). Energy production from biogas: A conceptual review for use in Nigeria. *Renewable and Sustainable Energy Reviews*, 32, 770–776. doi: 10.1016/j.rser.2013.12.013.

Schmidt, O. et al. (2017). Future cost and performance of water electrolysis: An expert elicitation study. *International Journal of Hydrogen Energy*, 42(52), 30470–30492.

Sebbahi, S., Assila, A., Alaoui Belghiti, A., Laasri, S., Kaya, S., Hlil, E.K., Rachidi, S. Hajjaji, A. (2024). A comprehensive review of recent advances in alkaline water electrolysis for hydrogen production. *International Journal of Hydrogen Energy*, 82, 583––599. doi: 10.1016/j.ijhydene.2024.07.428.

Serik, A., Kuspanov, Z., Daulbayev, C. (2026). Cost-effective strategies and technologies for green hydrogen production. *Renewable and Sustainable Energy Reviews*, 226(Part A), 116242. doi: 10.1016/j.rser.2025.116242.

Shen, H., Crespo del Granado, P., Jorge, R.S., Löffler, K. (2024). Environmental and climate impacts of a large-scale deployment of green hydrogen in Europe. *Energy and Climate Change*, 5, 100133. doi: 10.1016/j.egycc.2024.100133.

Staffell, I., Scamman, D., Velazquez Abad, A., Balcombe, P., Dodds, P.E., Ekins, P., Shahd, N., Warda, K.R. (2019). The role of hydrogen and fuel cells in the global energy system. *Energy & Environmental Science*, 12(2), 463–491.

Tahmasbi, M., Siavashi, M., Ahmadi, R. (2025). A comprehensive review of hydrogen production and storage methods: Fundamentals, advances, and SWOT analysis. *Energy Conversion and Management: X*, 26, 101005. doi: 10.1016/j.ecmx.2025.101005.

Tang, O., Rehme, J., Cerin, P. (2022). Levelized cost of hydrogen for refueling stations with solar PV and wind in Sweden: On-grid or off-grid? *Energy*, 241, 122906. doi: 10.1016/j.energy.2021.122906.

Turner, J.A. (2004). A realizable renewable energy future. *Science*, 305(5686), 972–974.

UNIDO (2021). Green Hydrogen: Enabling the Energy Transition in Developing Countries. UNIDO.

Wang, S., Liu, X., Zhou, C., Hu, J., Ou, J. (2017). Examining the impacts of socioeconomic factors, urban form, and transportation networks on CO2 emissions in China's megacities. *Applied Energy*, 185(Part 1), 189–200. doi: 10.1016/j.apenergy.2016.10.052.

Xu, Y., Cai, S., Chi, B., Tu, Z. (2024). Technological limitations and recent developments in a solid oxide electrolyzer cell: A review. *International Journal of Hydrogen Energy*, 50(Part C), 548–591. doi: 10.1016/j.ijhydene.2023.08.314.

Yang, C., Dong, L., Gao, Y., Jia, P., Diao, Q. (2021). Engineering acetogens for biofuel production: From cellular biology to process improvement. *Renewable and Sustainable Energy Reviews*, 151, 111563. doi: 10.1016/j.rser.2021.111563.

Zeng, K. and Zhang, D. (2010). Recent progress in alkaline water electrolysis for hydrogen production and applications. *Progress in Energy and Combustion Science*, 36(3), 307–326.

List of Authors

Sivathanu Pillai A
Adjunct Faculty
IIT Madras
India

Almas
Rajiv Gandhi Institute of Petroleum Technology
Bengaluru Campus
Karnataka
India

Venkata Pavan Kumar ARAPIRALA
Department of Electrical and Electronics Engineering
Madanapalle Institute of Technology and Science
Andhra Pradesh
India

Venkatesh BODDAPATI
Department of Electrical and Electronics Engineering
B.M.S. College of Engineering
Bangalore
India

Prakash D.B.
Department of Electrical and Electronics Engineering
B.M.S. College of Engineering
Bangalore
India

Binu Ben Jose D.R.
School of Electrical Engineering
Vellore Institute of Technology
Chennai
India

Biswanath DEKARAJA
Department of Electrical Engineering
AEC Guwahati
Assam
India

Arun J
Department of Chemical Engineering
KPR Institute of Engineering and Technology
Arasur
Coimbatore
India

Subramani K
Department of Physics
PSNA College of Engineering and Technology (Autonomous)
Tamil Nadu
India

Balasabarinathan K.V.
Department of Chemical Engineering
KPR Institute of Engineering and Technology
Arasur
Coimbatore
India

Ravulakari Kalyan
Department of Electrical and Electronics Engineering
B.M.S. College of Engineering
Bangalore
India

Pranitha Karthikeyan
IITM Pravartak Technologies
Chennai
India

Ilampoornan M.K.
Department of Electrical and Electronics Engineering
Bharath Institute of Higher Education and Research
Chennai
India

Chakkarapani Manickam
RGIPT Sivasagar Campus
Assam
India

Venkatesh Murugan
Department of Electrical and Electronics Engineering
Madanapalle Institute of Technology and Science
Andhra Pradesh
India

Arunkumar N
Department of Chemical Engineering
KPR Institute of Engineering and Technology
Arasur
Coimbatore
India

Kumaresan N
Department of Electrical and Electronics Engineering
B.M.S. College of Engineering
Bangalore
and
Department of Electrical and Electronics Engineering
National Institute of Technology
Tamil Nadu
India

Suriya Ponnambalam
Department of Electrical and Electronics Engineering
Bharath Institute of Higher Education and Research
Chennai
India

Venkatesh R
Department of Physics
PSNA College of Engineering and Technology (Autonomous)
Tamil Nadu
India

Satish Kumar RAMOJI
Department of Electrical and Electronics Engineering
Madanapalle Institute of Technology and Science
Andhra Pradesh
India

Arul Daniel S.
Department of Electrical and Electronics Engineering
National Institute of Technology
Tamil Nadu
India

Manimegalai S.
School of Electrical Engineering
Vellore Institute of Technology
Chennai
India

Nethravathi S
Department of Electrical and Electronics Engineering
B.M.S. College of Engineering
Bangalore
India

Lalit Chandra SAIKIA
Department of Electrical Engineering
NIT Silchar
Assam
India

Nindra SEKHAR
Department of Electrical and Electronics Engineering
B.M.S. College of Engineering
Bangalore
India

Sivasankari SUNDARAM
Rajiv Gandhi Institute of Petroleum Technology
Bengaluru Campus
Karnataka
India

Venkata Madhavaram T
Department of Electrical and Electronics Engineering
B.M.S. College of Engineering
Bangalore
India

Index

A, B

C, D

E, F

G, H

I, L

M, N

P, R

S, T

U, V, W

Other titles from

iSTE

in

Energy

2026

DOBIGNY Laure, ROBYNS Benoît
Rethinking Networks in Times of Transition: Ecological, Energy and Social Issues

2024

BENALLOU Abdelhanine
Heat Exchangers: Design and Sizing Algorithms
(Energy Engineering Set – Volume 6)

FAVENNEC Jean-Pierre
Geopolitics and Energy Transition 1: The Basics
Geopolitics and Energy Transition 2: From One Continent to Another, Contrasting Situations

MORY Mathieu
Mechanical Energy Conversion: Exercises for Scaling Renewable Energy Systems

ROBYNS Benoît, DOBIGNY Laure, ABBES Dhaker, DURILLON Benoit, BARRY Hervé, SAUDEMONT Christophe
Smart Grids and Buildings for Energy and Societal Transition

2023

ALVAREZ-HERAULT Marie-Cécile, GOUIN Victor, CHARDIN-SEGUI Trinidad, MALOT Alain, COIGNARD Jonathan, RAISON Bertrand, COULET Jérôme
Distribution System Planning: Evolution of Methodologies and Digital Tools for Energy Transition

BEN KHADER BOUZID Monia, CHAMPENOIS Gérard
Electrical and Mechanical Fault Diagnosis in Wind Energy Conversion Systems

JANNOT Yves, MOYNE Christian, DEGIOVANNI Alain
Heat Transfer 1: Conduction and Convection
Heat Transfer 2: Radiation and Exchangers

ROBYNS Benoît, LENGLET Claude, BARRY Hervé, BOZZO-REY Malik
Smart Users for Energy and Societal Transition

2022

ABOUELATTA Mohamed, SHAKER Ahmed, GONTRAND Christian
Smart Power Integration

BOUTAUD Benoit
Energy Autonomy: From the Notion to the Concepts

GONTRAND Christian
Electromagnetism: Links to Special Relativity

2021

ROBYNS Benoît, DAVIGNY Arnaud, FRANÇOIS Bruno, HENNETON Antoine, SPROOTEN Jonathan
Electricity Production from Renewable Energies (2nd edition)

2020

BOISGIBAULT Louis, AL KABBANI Fahad
Energy Transition in Metropolises, Rural Areas and Deserts

SOUALHI Abdenour, RAZIK Hubert
Electrical Systems 1: From Diagnosis to Prognosis
Electrical Systems 2: From Diagnosis to Prognosis

2019

BENALLOU Abdelhanine
Energy Transfers by Convection
(Energy Engineering Set – Volume 3)
Energy Transfers by Radiation
(Energy Engineering Set – Volume 4)
Mass Transfers and Physical Data Estimation
(Energy Engineering Set – Volume 5)

LACHAL Bernard
Energy Transition

2018

BENALLOU Abdelhanine
Energy and Mass Transfers: Balance Sheet Approach and Basic Concepts
(Energy Engineering Set – Volume 1)
Energy Transfers by Conduction
(Energy Engineering Set – Volume 2)

JEMEÏ Samir
Hybridization, Diagnostic and Prognostic of Proton Exchange Membrane Fuel Cells: Durability and Reliability

RUFFINE Livio, BROSETA Daniel, DESMEDT Arnaud
Gas Hydrates 2: Geoscience Issues and Potential Industrial Applications

VIALLET Virginie, FLEUTOT Benoit
Inorganic Massive Batteries
(Energy Storage – Batteries and Supercapacitors Set – Volume 4)

2017

BROSETA Daniel, RUFFINE Livio, DESMEDT Arnaud
Gas Hydrates 1: Fundamentals, Characterization and Modeling

LA SCALA Massimo
From Smart Grids to Smart Cities: New Challenges in Optimizing Energy Grids
(Advanced Smart Grids Set – Volume 2)

MOLINA Géraldine, MUSY Marjorie, LEFRANC Margot
Building Professionals Facing the Energy Efficiency Challenge

SIMON Patrice, BROUSSE Thierry, FAVIER Frédéric
Supercapacitors Based on Carbon or Pseudocapacitive Materials
(Energy Storage – Batteries and Supercapacitors Set – Volume 3)

2016

ALLARD Bruno
Power Systems-on-Chip: Practical Aspects of Design

ANDRÉ Michel, SAMARAS Zissis
Energy and Environment
(Research for Innovative Transports Set – Volume 1)

DUFOUR Anthony
Thermochemical Conversion of Biomass for the Production of Energy and Chemicals

SOUSTELLE Michel
Phase Transformations
(Chemical Thermodynamics Set – Volume 5)
Thermodynamics of Surfaces and Capillary Systems
(Chemical Thermodynamics Set – Volume 7)

2015

MONCONDUIT Laure, CROGUENNEC Laurence, DEDRYVÈRE Rémi
Electrodes for Li-ion Batteries: Materials, Mechanisms and Performance
(Energy Storage – Batteries and Supercapacitors Set – Volume 2)

ROBYNS Benoît, FRANÇOIS Bruno, DELILLE Gauthier, SAUDEMONT Christophe
Energy Storage in Electric Power Grids

ROSSI Carole
Al-based Energetic Nanomaterials: Design, Manufacturing, Properties and Applications
(Nanotechnologies for Energy Recovery Set – Volume 2)

SOUSTELLE Michel
Phases Modeling Tools: Application to Gases
(Chemical Thermodynamics Set – Volume 1)
Modeling of Liquid Phases
(Chemical Thermodynamics Set – Volume 2)
Thermodynamic Modeling of Solid Phases
(Chemical Thermodynamics Set – Volume 3)
Chemical Equilibria
(Chemical Thermodynamics Set – Volume 4)
Ionic and Electrochemical Equilibria
(Chemical Thermodynamics Set – Volume 6)

TARASCON Jean-Marie, SIMON Patrice
Electrochemical Energy Storage
(Energy Storage – Batteries and Supercapacitors Set – Volume 1)

2014

DE LARMINAT Philippe
Climate Change: Identification and Projections

2012

BECKERS Benoit
Solar Energy at Urban Scale

FOULADGER Javad
Electrothermics

2011

GAO Fei, BLUNIER Benjamin, MIRAOUI Abdellatif
Proton Exchange Membrane Fuel Cell Modeling

MONMASSON Eric
Power Electronic Converters: PWM Strategies and Current Control Techniques

MULTON Bernard
Marine Renewable Energy Handbook

RAZIK Hubert
Handbook of Asynchronous Machine with Variable Speed

2010

BRUNET Yves
Energy Storage

JUFER Marcel
Electric Drives: Design Methodology

2009

SABONNADIÈRE Jean-Claude
Low Emission Power Generation Technologies and Energy Management
Renewable Energy Technologies

2008

MEUNIER Gérard
The Finite Element Method for Electromagnetic Modeling

Printed and bound by CPI Group (UK) Ltd, Croydon, CR0 4YY

26/08/2026

14949808-0001